C0-AJQ-604

QUALITY IS EVERYBODY'S BUSINESS

QUALITY IS EVERYBODY'S BUSINESS

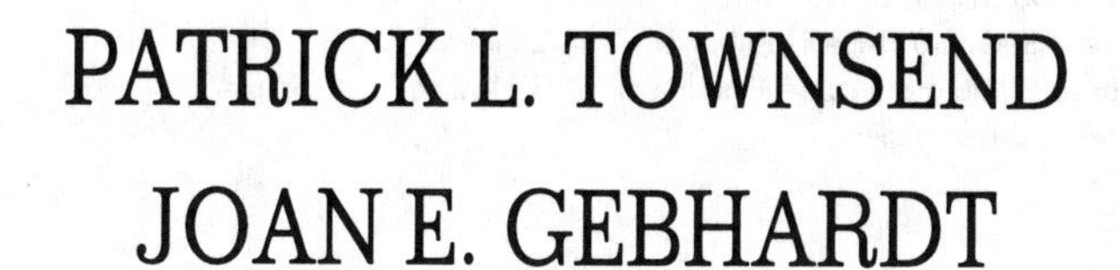

PATRICK L. TOWNSEND

JOAN E. GEBHARDT

St. Lucie Press

Boca Raton London New York Washington, D.C.

Catalog record is available from the Library of Congress

International Standard Book Number 1-57444-284-8
Printed in the United States of America 1 2 3 4 5 6 7 8 9 0
Printed on acid-free paper

INTRODUCTION

There are four compelling reasons for an organization to make the investment and take on the tasks necessary to define, initiate, and maintain a robust quality process:

- Done correctly, quality makes a great deal of money and/or conserves an enormous number of resources.
- Done correctly, quality attracts and retains customers.
- Done correctly, quality makes employees happy.
- Done correctly, quality is the ethical thing to do.

The key, obviously, is to "do" quality correctly. Too few organizations meet the test.

Unfortunately, when quality efforts fail to fulfill their potential, business leaders begin to doubt the efficacy of making the pursuit of quality a primary organizational priority.

A consistent mistake has been to "start small" and implement only a portion of what should be done. Examples of partial efforts ending in disappointment or disaster abound, with the most obvious and most painful examples being reengineering programs that focused solely on "Doing the right things" and ignored "Doing things right." As a result, the only thing "total" about most Total Quality Management processes has been the level of frustration experienced by employees throughout the company.

The objective of this book is to make it possible for people at all levels of an organization to understand both the underlying theory and the specific mechanics of what the authors have come to believe are the essentials of quality.

The book consists of 49 articles published individually over a period of 14 years. The principles have remained consistent while the vocabulary and recommended tools have grown and evolved as the authors' experience and knowledge of the subtleties of quality increased.

Taken as a whole, these articles address both the theory and practice of quality in an integrated manner that makes it possible for an organization to understand what must be done and why, as well as what can be reasonably expected as a result of their investment.

Perhaps the most important single fact that senior managers must grasp if their organizations are to survive in a competitive environment is that "quality" is not going to go away. While some business executives have tried to treat quality as a transitory concept, customers do not consider quality to be a fad.

Educated customers are quality's best friend. As customers learn more and more about their options, the demand for quality increases. And once a customer experiences quality in the product and/or service that he or she receives, the customer will always look around to find the quality option. Ironically, many of the same executives who pay little attention to quality in their own organizations are the same people who are extremely demanding in their roles as consumers of goods and services.

Quality is not easy. It is not particularly complex, but it is hard to do because continual attention is required to insure continual improvement. Done correctly (ahh, there's that phrase again), a quality process can be defined and implemented in an organization of 3000 people or fewer in six to eight months—and that includes actively involving every person on the payroll in the process and seeing positive bottom-line impact virtually immediately.

Enough said. Whether an organization has started a quality process previously or it is now considering a first attempt, as the title of the first article in the first chapter of the first section of this book puts it, "It Is Time to Get on with It!"

ABOUT THE AUTHORS

Patrick Townsend and Joan Gebhardt—the founders and principals of Townsend & Gebhardt—have given presentations and conducted workshops throughout the world on the topics of quality and leadership. Among the ideas that have established their credentials as being uniquely valuable to audiences in India, the United States, South America, and Europe are their emphasis on the importance of 100 percent employee involvement; their explanation of how to establish—within a matter of months—a quality process that brings benefits to owners, customers, and employees; and their introduction of the idea that a variety of pragmatic and effective leadership lessons can be drawn from military sources for application in the business world.

For instance, they believe that quality is worth pursuing because:

- Done correctly, it makes a great deal of money and preserves resources.
- Done correctly, it makes customers happy.
- Done correctly, it makes employees happy.
- Done correctly, it is the ethical thing to do.

The authors can be contacted at:
townsendgebhardt@worldnet.att.net

CONTENTS

SECTION FIVE: FINAL THOUGHTS [illegible]

1

ORIGINS

Call to Action!
Democracy and Quality

The first two chapters — and seven articles — describe reasons to pursue quality, address the functional differences between quality in a service organization and quality in a manufacturing concern, and discuss the important but little-explored link between a quality process and the political climate of the nation.

By giving theory, motivation, and specific examples, the section provides background for the remainder of the book.

1

CALL TO ACTION!

In "It Is Time to Get on With It!," the underlying premise is that "For a top executive... not to acknowledge that a pursuit of quality... is financially beneficial requires an act of willful indifference to the data available." The article goes on to address the most-often heard excuses for putting off a quality effort until "the time is right" and proposes that the best time to begin is immediately. The longer a company puts off its inevitable quality effort (made inevitable by the demands of customers), the greater the advantage they cede to their competitors.

"The Service Revolution" and "A New Model for Quality" were written when the Quality Revolution in the United States was just beginning to have an impact on the service sector of the economy. Interestingly, the proposition that quality should (and could) be pursued in manufacturing was easily accepted many years before the idea of implementing a formal quality process in a service company was seriously considered by many people. Looking back, the notion that quality might not be appropriate in service organizations seems a bit bizarre—especially considering Dr. W. Edwards Deming's estimate that, in the late 1980s, 86% of American workers worked in service-oriented jobs (either in service companies or in the service departments of manufacturing companies).

The United States is not alone in this bias toward service. A 1993 study indicated that the service sector of the Mexican economy accounted for 63 percent of the gross national product and more than 50 percent of the employment. Both figures were expected to rise as the economy matures, in keeping with the historical shift from manufacturing efforts to service industries which usually occurs in maturing economies. In discussing the inherent differences between quality efforts in service organizations and those in manufacturing organizations, "Total Service Quality" offers insights into the change in techniques necessary when going from one to the other.

In both "The Service Revolution" and "A New Model for Quality," the Paul Revere Insurance Group, one of the first service companies to fully embrace the idea of a full-fledged, organization-wide pursuit of quality, is cited. The Paul Revere "Quality Has Value" process appears as a recurring example throughout this book, and Chapter 11 offers a thorough description of the process's components.

IT IS TIME TO GET ON WITH IT!

The time for exhaustive studies and pilot programs is past. For a top executive in America not to acknowledge that the pursuit of quality – in particular a participative quality process that actively involves 100 percent of the men and women on the payroll – is financially beneficial requires an act of willful indifference to the data available. Quality, in fact, may be the key to the survival of any organization into the 21st century.

For years, many executives hid behind the insulting (to their own employees) idea that, "Well, this quality push seems to work for the Japanese, but American employees are different. They won't work as hard; they change jobs more frequently; they just aren't as interested." But now there are too many American quality success stories (from companies of all types and sizes) to ignore.

The Malcolm Baldrige National Award process, for instance, has named [as of late 1989] five splendid examples from the manufacturing segment of the American economy. Four of the winners have been large organizations, but one of them, Globe Metallurgical Incorporated, competes in the international market-place with fewer than 500 employees. Equally valuable examples from the largest segment of the economy, service, are also available. Three of these, Paul Revere Insurance Group, L.L. Bean, and USAA Insurance, are mid-size companies.

LEADERSHIP AND EXCUSES FOR NOT GETTING STARTED

Quality and the competitive benefits it invariably brings are within the reach of any organization whose leadership is willing to follow the proven, if not yet completely, beaten-flat, path blazed by these pioneers. In spite of this, executives find numerous, and ostensibly powerful, excuses for inaction when it comes to quality.

Can't start until it's perfect The first of these might be deemed "the search for the perfect progress." This often begins with the "program versus process" debate. Practitioners in the field of quality are careful to point out the difference between the two. A program, it is explained, is something with a definable beginning and a pre-arranged ending; it is short-term in its nature. A process, on the other hand, has a definite beginning but it has no end; it represents a change in the way things are done forever. Quality, it is accurately proclaimed, requires a process.

For many executives this supplies a dead-earnest, cast-iron excuse for making a decision not to try anything. While it is true that for most companies the beginning of a quality process represents a major cultural change, it is not true that because a process is going to last an indefinitely long period of time, it should take an indefinitely long period of time to begin. A quality process is not something to be undertaken lightly, but it is possible to begin modestly, to be right without being perfect. Trade-offs and compromises can be made.

Leaders don't have time; aren't charismatic There is only one point on which there can be no compromise. To get anything done requires top management commitment – the consensus "most valuable concept" of quality. Without the informed, active, and obvious belief and commitment of the leaders of the organization, any efforts at quality will be foredoomed to fall far short of potential.

This provides another excuse for some executives who feel they lack the time or the charisma to make such a commitment. Time is, indeed, essential. Charisma, however, is not. Top executives need to give up a few hours every week for several months while hammering out exactly how this quality process is going to work in their unique organization. They need to give up time for their personal, ongoing participation in whatever process they define, both by improving quality in their areas and by recognizing the efforts of others. They need to commit the time of others and provide the financial backing for training and other support programs.

One way executives can leverage their time is with the use of consultants. In the early stages of defining a process, one or more consultants can be invaluable sources of ideas and perspective. Later on, they can initiate training. Still later, they can be used for re-evaluation.

Leadership and consultant traps to avoid Even with the use of consultants, however, the establishment of a quality process cannot be immediate; no off-the-shelf approach can be purchased on Friday for "installation over the weekend." But neither should it take forever. Too many attempts to begin quality processes turn into lifetime annuities for consultants.

If the relationship between a company and a consultant hired to assist in the establishment of quality process drags open-ended beyond a year, the worth of the contribution of the consultant and/or the depth of the commitment of top management of the organization is open to serious doubt.

If, after all, top management is serious, a year is more than enough time to hire a permanent staff. And if the consultant did not begin—shortly after arrival—to lay the groundwork for his or her own departure, the company hired the wrong consultant.

Be very specific when hiring consultants... A rule of thumb is to hire a consultant for a specific skill for a specified period of time for an agreed-upon price—and with the provision/deal that all knowledge and materials are left behind. Independence from future obligations to a consultant is worth a reasonable price for reproduction rights (for internal use only) for tapes, books and such. If the first consultant spoken to isn't comfortable with these stipulations, a company can rest assured that somewhere there is a consultant, who is at least as good, who is.

Moving direction and training in-house As quickly as possible, all training and direction needs to be brought in-house—so that there is no doubt in any employee's mind about whose process this is. The building of a permanent structure, directed by a staff of full-time employees is one of the more obvious signs of commitment by top management. As long as top management retains a framework of consultants who are the "real experts," everyone knows that this quality process could end tomorrow. Consultants can be let go almost as easily as they are hired. A consultant-driven quality process always looks suspiciously temporary.

WHAT DOES TRUST HAVE TO DO WITH COMMITMENT?

Top management commitment may not require charisma, but it does depend upon a far more accessible trait—trust. This translates into acceptance on the part of top management that the payroll is made up of adults.

Better yet, executives need to openly admit that the adults on the payroll know their jobs at least as well as they have been allowed to learn them and that they want the organization to not only survive, but succeed.

Strangely, top management often entrusts executives with sensitive information and projects, but rarely does the same with hourly workers. Yet, the chances of an hourly worker getting a call from a head-hunter in mid-project and leaving with all his or her newly acquired knowledge are far lower than the odds of that same thing happening to an on-the-rise executive. If anything, the folks on the lower end of the corporate ladder are often more bound to, and more ego-involved in, the organization than the folks from the middle on up. The continued success of the company is not optional for them.

What logically follows the acceptance that the overwhelming majority of the people on the payroll are adults are questions concerning how to involve everyone in improving the operation of the organization. Too often these logistics of participation provide additional excuses for inaction. These fall into several categories.

WHO'S IN THE QUALITY PROCESS?

Often, the first question asked is, "Which of our people should we include in this quality process?" It's a no-win question because someone will be left out. A better question for the leadership of an organization to ask at the outset of a quality process is: "Who can we afford to exclude from our effort to improve, i.e., who do we think can never get any better than they are today?" The answer had better be "Nobody."

"Pilot programs" then take on a different aspect. They are a beginning of a process that will, according to an announced schedule, involve everyone. The expansion of pilot programs must then be relatively quick and highly visible. To limit a quality effort to a minority of the workforce through the use of a few pilot programs stretched out over endless months sends a very clear message to the majority of the people on the payroll:

We [*read, top management*] are reluctant to let employees address quality issues.

Logistical problem: training everyone Another logistical problem of 100 percent participation concerns training. Unfortunately, after deciding to include everyone, some organizations allow themselves to get trapped in a meaningless training cycle.

Convinced (quite often with the help of a consultant/training company with "just the right" training package) that they can't begin until every employee has been through a multi-day training program, companies begin training their way through the workforce without providing a structure that makes it possible for the training to be immediately applied.

A couple of years after beginning such an exhaustive, and exhausting, training program, when the training department comes up for air they find out that the people who went through the initial courses either are no longer with the company (having gotten tired of waiting) or are no longer particularly interested (it's tough to keep enthusiasm up in a vacuum).

Getting the ball rolling One structure that can be used during this ongoing training is quality teams. The training can be used first to create team leaders who have a background in how to conduct a meeting in a participative, efficient manner, as well as some training in problem solving. Also, the quality team leaders should have a thorough grasp of the mechanics of the organization's quality process. This must be supplemented by a one- or two-hour-long (no more) briefing on the company's quality process and the role of the individual for all employees.

As time passes, more and more people can receive formal training on SPC and other helpful tools that will be of use as they move into addressing more and more complex problems. In fact, once launched, any successful quality process is the catalyst for all sorts of training as the process, and the organization, begins to evolve and change.

Yes, it would be wonderful if every person on the payroll were intimately familiar with all the nuances of the concept of quality and the Pareto technique (and every other SPC option), and if every team of workers had been fine-tuned and certified prior to the formal launch of the process. But it is far more important to get started and begin building an atmosphere of Kaizen, of continuous improvement. As that great American philosopher, Bo Jackson, has said, "Just do it."

ROLES FOR MANAGEMENT AND EMPLOYEES

One concept that will help to jump start a quality process is to address quality as having two distinct aspects: *doing the right things* and *doing things right.*

Choosing which are the right things to do is the job of management (i.e., the next level up) at each level. Deciding on how best to do those things—doing things right—falls most frequently into the province of the people who actually produce the agreed upon "right" product or service.

If quality teams (or whatever name is used to designate groups of employees at every level of an organization) are given the charge, "Do things better," work can start very quickly.

Employees can assume that what they've been told to do is the right thing to do (in fact, it may be under assessment by a quality team at a higher level or by a separate effort such as a value analysis workshop) and begin to explore better ways to do it. By not concentrating solely on a few big projects that will take months to study and then must be reviewed by the whole world before any action is approved, participants are freed to make small improvements, dozens of them.

The trade-offs between limiting quality improvement to small doses and getting everyone involved as quickly as possible heavily favor the latter approach.

Beware of a pilot under another name: executive task forces—there are advantages to beginning with just a small percentage of people and to confining efforts to a few "task forces." Relatively little up-front time is needed. Just pack a collection of executives off to one school or another and when they come back, they can agree to a few groupings of people (led by moderately high executives), attack a small number of projects (probably identified at the school they went to), turn the whole thing over to a few consultants, and get back to doing "real work" (properly exhilarated, of course).

After a year or so, another collection of people can be packed off to school, and the cycle can be repeated. This approach will pay for itself, won't be a big bother to anyone in the upper echelons of the company, will give the president of the company something to write about in the annual report, and will keep a few consultants from being a burden on local welfare rolls. The gains, however, will be relatively small and there will be little fresh ground broken since the action teams are limited by definition to confirming the ideas of their superiors.

THE WEALTH OF ORGANIZATIONS

To involve everyone, to take advantage of the tremendous wealth of knowledge already resident in the company, requires a broader approach and involves a risk. A structure will need to be defined and put in place. Appropriate training will have to be provided to all employees. Eventually, decision-making will have to be pushed down the corporate ladder. Power to make changes, long since usurped from subordinates, will need to be returned.

Rather than having a few carefully supervised projects grinding towards completion at the end of the first year, the company that takes this second approach will encourage hundreds, if not thousands, of improvement efforts, big and small. Perhaps the only drawback is that employees will expect to be taken seriously, which is not always convenient or comfortable.

The action and results that come with a 100 percent involvement process engenders far more more good will and participation than inspirational speeches, hours of training, and/or the thrill of watching a few pilot groups step through laborious, management-driven procedures. It is true that some mistakes may occur, but why should the organization cut itself off from the 99 percent of the ideas that are good just to avoid the 1 percent that are not?

WHEN CAN WE START?

The time to begin is now. The models exist. There can be no more excuses. The necessity to openly entrust the future of the organization to people other than the CEO and his or her direct reports is the key to quality and to competitive success in the 21st century.

Journal for Quality and Participation, *March 1990.*

THE SERVICE REVOLUTION

Andy Warhol is credited with saying that every person gets a fleeting pass at fame. The same may be said for words. Some words come into vogue and are then used by so many people in so many ways that they quickly lose any specific meaning. Which is why I am going to start by defining the word service. It has several distinct, yet overlapping meanings. *Service* is used to describe both a type of industry and a single business transaction in which the consumer is left with either a memory or a piece of paper when the transaction is complete.

This covers a wide range of possibilities from restaurants and hotels to insurance companies and banks. It also covers work done on a previously purchased manufactured item, both when the work being done is covered by a warranty or as a new expense.

What I will address here will be keyed in the main to the paper-and-ideas industries, such as insurance companies, software firms, and banks, although the underlying concepts and, indeed, many of the techniques are quite applicable in all aspects of business.

I have served as the director of two highly successful quality processes in service organizations. One, the Paul Revere Insurance Group, is what might be called "medium tech," while the other, McCormack & Dodge, is a high-tech organization that deals with software development and software services. My remarks will be drawn mostly from the Paul Revere experience, since it has been established the longest and thus offers the widest range of lessons. In fact, Paul Revere's quality process, formally launched on Friday, 13 January 1984, is one of the real innovators and pioneers in this field.

The distinguishing characteristics of these processes–and the primary lessons to be learned by any organization looking to the establishment of a quality process in the years to come—are: (1) the level of top management commitment and the ways it was actually practiced and (2) the formal involvement of every single person on the payroll in the effort of establishing a habit, or a culture, of continuous improvement.

I realize that everybody who does any proclaiming at all on the topic of quality processes lists top management commitment as the first, and most necessary, element. But let's look beyond the simple phrase for a few minutes. "Top management commitment" does not simply mean that the CEO—or president or whoever—must be willing to venture forth from his or her office once a year to deliver a rousing speech and also to sign the often outrageously large checks for a band of consultants who are supposed to be "installing" some sort of quality process.

Top management commitment must be informed, active, and obvious if it is to be believable and effective. The top executives must not simply allow the quality process to take place; they must be active in its definition and, once begun, in its implementation. Examples would include membership on the committee that actually defines the process that will be used in their organization, actively working to improve their own contribution to the organization, and taking part in the many occasions of recognition, gratitude, and celebration that are an integral part of a well constructed quality process.

At Paul Revere, the Quality Steering Committee was made up of the number-one or number-two executive from every division and major department of the company. The president and his staff made up a Quality Team known as the Big Guys, and their ideas and progress were listed on the Quality Team Tracking Program along with those of every other team in the company – available for all to see. So far as actively saying "Thank You" for contributions in the first four years of the process, the president of the company and his top executives took part in over 1,500 awards ceremonies.

Top management commitment also means accepting the idea that the personnel department has been hiring adults for all these years – and that those people, those non-management people on the payroll, can read and write, and they want very much for the company to succeed. This belief in the abilities and intentions of the employees is best demonstrated by pushing authority down the corporate ladder.

I am not advocating a loosey-goosey, wildly intrapreneurial operation. I am describing an ordered business process in which authority is equal to responsibility at the team level. If a group of employees–at any level–is on the line for something, if they are going to get blamed if it goes wrong, then they should be able to change it. Authority equal to responsibility–no more, but no less.

Knowing that an idea will be listened to, that changes are possible, encourages people who always before did all their creative thinking outside of their work place to begin focusing some of that talent on taking charge of their own work environment and improving their contribution to the final product or service. Companies that are to be successful in the years to come will be so because they are taking advantage of the skills of all of their people.

Which leads me to the other characteristic that I mentioned–100 percent involvement. I don't use the word "mandatory," but I do use "non-voluntary." Why should it be considered unusual or onerous to want everybody on the payroll to contribute to the continual improvement of the organization? Quality processes that begin by defining a specific subset of the employee base—either "volunteer only" or any other grouping—are limiting their success by definition. It would be like a runner deciding to run in the Olympics but then deciding to use only one leg during the 100 meter dash.

But 100 percent employee involvement, what my partner and I call Participative Quality, can't just be wished for. There must be a specific structure established that makes it possible for every employee at every level to take an active role and that, in fact, makes it obvious if they don't.

At Paul Revere and at McCormack & Dodge, every single person on the payroll is on a Quality Team. The teams are composed of about ten people each and are encouraged to meet weekly for about 30 minutes. The Quality Team Leaders are trained in how to run a meeting in a participative manner and how to solve problems using a variety of tools. The teams are told that they have authority equal to their responsibility, and they are encouraged to improve their business processes.

In the abstract of Paul Revere's application for the Malcolm Baldrige National Quality Award last year, (incidentally, they were one of two service finalists, although no awards were given in the service category) it stated that a person could walk into Paul Revere, walk up to any employee, and ask what Quality Team the person was on and what he or she personally was doing that week to improve the quality of the organization, and get specific answers to both questions. That's what I mean by 100 percent involvement.

Be warned, however. This quality stuff is not easy, it is not quick, and it is not free. It is, however, necessary for any company that wants to enter the next century as an independent corporate entity, much less thrive in the 21st century.

The simple fact that a company has survived until now does not guarantee its continued existence. Everyone's performance hasn't suddenly gotten worse. What has happened in many industries, both service and manufacturing, is that standards have been raised by the better, more reliable products and services being offered by some organizations in the marketplace–where virtually every commercial outlet in America carries a wide selection of products and services of non-American origin.

The level of dissatisfaction evident among American consumers comes from their growing realization that the abuse they have experienced for so long at the hands of

American businesses didn't have to happen. It is a bit like the old joke about no one realizing that a mutual acquaintance drank until he was seen sober one day. Consumers didn't know the products and services they were buying were "bad" until a few companies began treating them well. Not by accident, those companies have reaped great profits.

And it is going to get even more competitive. In the past, the question was: "Who's good enough?" Currently, the question is: "Who's getting better?" In the future, the question will be: "Who's getting better faster?"

It won't be easy to keep pace. But, while the bad news may be that quality is hard to come by and tough to maintain, the good news is that it is also the surest way to financial success. Quite simply, there is only one real reason to go through the pain of establishing a quality process: it makes a lot of money. On a grander scale, the countries whose companies understand and act upon the principles of quality—in service and manufacturing—will, through their economic dominance, define the future.

The evolution of a nation from a manufacturing-based economy, or from an economy with a healthy balance between manufacturing and service, into a service-dominated economy does not necessarily represent a maturing of the economy. In the case of the United States, it means only that our manufacturing enterprises have been consistently beaten and that as a national economy, we have retreated to service.

What has not been possible in the past through military means is now possible through economic competition. Business beachheads are, quite simply, tougher to repel than a force of Marines with weapons blazing. For one thing, business commandoes are less obvious; for another, they often bring very real immediate benefits to the "invaded" country—such as jobs.

Ironically, there really isn't much constructive that the government can do about it. Tom Peters estimates that one half of one percent of the current U.S. trade deficit can be traced to trade policies. The rest is due to the fact that US manufacturers–and increasingly, service providers–don't offer products or services that work the way they are advertised to work or the way buyers want them to work. Already, the United States has an international trade deficit in service adding to its huge manufacturing deficit.

The government can help with supportive measures and education, such as provided by the Malcolm Baldrige National Quality Award, which was initiated in 1988. Business, however, has the opportunity to solve the challenge just waiting to be recognized. What business must do is abandon the search for the one magic consultant with the perfect system and dig in to do the hard work, which is necessary in order to enlist the brain-power already on the payrolls.

This effort will not be a genuine or gentle journey for most organizations. The word "journey," after all, conjures up the image of something done voluntarily or at a leisurely pace. While it is a process with no neat, "we're there, we're finished" ending, most organizations cannot afford to go slowly.

Thomas Jefferson once said that every generation needs a revolution. This generation's is in the workplace. A culture-wrenching revolution, led by pragmatic business leaders, is needed if companies and, indeed, countries are to retain their independence in the years to come.

Distinguished papers, *December 1989.*

A NEW MODEL FOR QUALITY

"How can we achieve quality?" has become the hottest question in American business during the last several years, equaled only by "How can we make more money?" In truth, if an organization can find the answer to the first question, it will have the answer to the second.

The launching of quality process is a hard-nosed business decision. A successful quality process will increase employee morale, make customers happy, boost job satisfaction, and lower turnover rates, but the reason a quality process is worth the effort is this: It makes a lot of money.

The problem comes in figuring out how to go about "doing quality," especially since the models being offered in most books and by an overwhelming percentage of consulting firms are rooted in manufacturing. Why follow a failed model? America's disastrous performance in the manufacturing world during the last few decades did not occur because there were no attempts at quality. The problems were caused in large part by a basic misunderstanding of the American work force by those who appointed themselves leaders of the quality movement.

If controls and specific goals defined by top management didn't produce quality, the experts insisted that more controls and more specific goals were needed. The problem, they proclaimed, was that the rebellious, under-educated, under-motivated American workers didn't follow instructions well. So the instructions were made more rigid, and the volume was turned up.

The fact that little got better didn't alter the thinking. Now the American economy finds itself dominated by the service industries, and the experts who contributed to the fall of manufacturing are busily telling the service industries what they should do to improve. Following the lead of the manufacturing-based quality and productivity experts is a ticket to disaster that the American economy cannot afford for its service or high-tech elements to buy.

If there is blame to be assessed for the failed manufacturing quality model, some must be given to the quality control experts and their myopic view of their role, as well as the academics who provided them with support. But the heaviest piece of the blame must go to the men and women at the top, the ones who hired and believed. By not knowing the strengths of their own workers, by not investing in their own plants, by not trusting either their own instincts or the people on their payroll, top management sealed the fates of their companies.

American companies that wish to survive to see the 21st century dare not ignore the challenge of quality, nor do they dare follow a failed model. This is not to say that such esteemed figures as Dr. W. Edwards Deming should be ignored–only that the bastardization of his management philosophies should be discarded. Deming's Fourteen Points included no requirements to take particular measurements at particular times or to set specific numerical goals for every aspect of operations.

The points are management principles that assume that appropriate measurements will be taken and used at appropriate times and levels. To do so will require trusting a lot of people with a lot of information–and sharing the power that manufacturing's top executives and quality control honchos have hoarded.

Tom Peters once wrote that the key to quality was respect–respect for the customer and respect for the employee. It really is that simple, and that difficult. To show respect for employees implies treating them as contributing, knowledgeable, capable human beings. It means making it possible for them to take a greater degree of control over their own workplace and to be active participants in the continual improvement of the organization. It means returning to the proper level the power that has been usurped from subordinates throughout American industry.

There is a developing model that appears to be working extremely well in both medium-tech and high-tech companies in the Bay State. At the Paul Revere Insurance Group, headquartered in Worcester, it is called "Quality Has Value," while at Natick-based McCormack and Dodge, it is called "Quality Without Limits." The chief characteristic of these processes is that everyone is involved, is formally enrolled.

At first glance, the 100 percent enrollment of employees on quality teams would seem more controlling than the traditional "quality control" checkers who sit by the door and attempt to stop any "bad stuff" from leaving the building—or having a dozen or fewer "quality circles" in an organization. However, it is precisely because the number of teams is so high that the power must be shared. It is easy to micro-manage a select few quality control checkers or to keep a close eye on the work of a defined (small) number of quality circles. But if top management has the courage to say to its employees, "You have the authority equal to your responsibility. Make things better," power is dispersed.

In the first 3½ years of Quality Has Value process, the 250 Paul Revere quality teams implemented over 25,000 quality ideas. The cost to produce a single insurance policy decreased steadily while customer satisfaction and sales headed up. At McCormack & Dodge, in the first four months of the Quality Without Limits process, the 150 quality teams logged over 1,700 quality ideas onto their "Quality Tracking System," with certified savings already in the millions.

While no quality process can be transferred without alterations to a second organization, there are certain principles that can and should be adopted. These principles include: (1) informed, involved, active commitment from top management (signing the checks for a consultant to "implement the process" and making an annual speech is not enough); (2) 100 percent involvement of everyone on the payroll, from the CEO to the high school dropout who was hired yesterday—(who are you willing to look in the eye and tell, "We don't care if you don't improve because you are not particularly important and/or clever"?) (3) the granting of authority commensurate with responsibility to every employee team; (4) a formal structure to continually remind employees of the process, and to give the process stability and permanence; and (5) a varied and vigorous program of recognition, gratitude, and celebration.

America's last hope to control its own economic destiny lies in the hands of the leaders of the service and high-tech segments of business. Those leaders must turn their backs on the manufacturing model and begin immediately to enlist all of their employees in this struggle. Quality is no longer an optional competitive tool; it is the key to survival.

Mass High Tech, *Oct 10th–Oct 23rd, 1988.*

2

DEMOCRACY AND QUALITY

Quality processes have a far higher chance of success in a democracy than under an authoritarian regime because of a crucial element that democracy and a well-conceived quality process share: both believe in the value of the individual. In a democracy, every individual has a vote based on the assumption that every individual has a right to have a say about his or her daily affairs; in a quality process, that same belief is the justification for opening the process to the involvement of every person on the payroll. An attempt to establish a quality process—particularly one with 100% employee involvement—in an authoritarian political setting would be extremely difficult at best. It might even be thought of as subversive.

These articles—beginning with "A Revolutionary Example for Quality"—were written for an American audience but international readers can easily substitute heroes of the birth of democracy in their country for the American historical characters mentioned: George Washington (the first president of the U.S.), Thomas Jefferson (author of the American Declaration of Independence and the 3rd president), Ben Franklin (politician, inventor, and diplomat), John Adams (the 2nd president), and Alexander Hamilton (politician and financial expert).

The statement by Konosuke Matsushita included in "An Excellent Enterprise if You Can Keep It" received wide circulation in the United States and served as a rallying point for American executives for years. It can be safely said that Mr. Matsushita played a major role in energizing the American Quality Revolution.

"The Revolution Continues" introduces the idea that a quality process has three major components: Leadership, Participation, and Measurement. It is a concept which first appeared in *Quality in Action: 93 Lessons in Leadership, Participation and Measurement* by these authors. It is further explored at several points in this book, beginning in the next section.

A REVOLUTIONARY EXAMPLE FOR QUALITY

It has long been fashionable to credit Drs. W. Edwards Deming, Joseph M. Juran, Tom Peters and other 20th century leaders for the quality revolution.

It is just as accurate, and perhaps more revealing, to give the credit to George Washington, Thomas Jefferson, and Ben Franklin, the 18th century leaders of the American Revolution. The same impulses informed both revolutions.

Both the American Revolution and the quality revolution hold the belief that institutions must be designed to meet the needs of people, rather than that people exist to perpetuate institutions. There is in both an equally strong belief that, over the long haul, decisions made by many people are apt to be wiser, kinder, and more rooted in reality than decisions made by a chosen few. In short, both revolutions have respect for the individual, what might be called the democratic ideal, as their touchstone.

America, however, has never been a pure democracy. It is not one today. When asked by a woman standing outside the Constitutional Convention what form of government had been designed for the newly formed United States, Ben Franklin replied, "A republic if you can keep it."

Imbedded in that reply is a great deal that Americans hold dear: separation of powers, the rule of law, responsibility as well as rights. For over two hundred years, the citizens of the United States have struggled successfully to keep their Republic, making government under the Constitution of the United States one of the oldest in the world. If the viability of a written constitution is used as the measure of longevity, the United States is currently the second oldest nation on earth, following Iceland.

During this period, the democratic ideal continued to evolve in the United States. In the political arena, the expanded right to vote is its hallmark. That franchise was first enlarged to include black men, then women of all backgrounds. Most recently, the minimum age was moved downward from 21 to 18. Throughout, the effort has been both to extend to people what is believed to be rightfully theirs and to create the situation in which the nation can benefit from the input of more people, more points of view.

In the workplace, the quality movement of the last two decades has sought to embody this same ideal: the principle that everyone is capable of contributing to the common good. But in the workplace, as in the political arena, a pure democracy would be chaotic at best. Looking at successful quality processes, processes with longevity, success is more likely where the democratic ideal has been tempered with the principles of a republic.

The distinguishing characteristic of a republic is the idea of representative government. The citizenry votes for representatives, who in turn vote on specific matters such as wars and budgets and welfare programs. Responsibility varies at each rung of the governmental hierarchy, both vertically and horizontally.

The principle of separation of powers into judicial, legislative, and executive may seem clumsy and/or slow at times but it has been a source of reasoned strength far more often than not.

So too in a private company; only in the rarest of situations is any decision put to an all-hands vote.

The sinew which strengthens to governance: national and corporate What makes the American government work — and what makes a successful quality effort work—is the recognition of rules and roles that gives the organization its cohesiveness. In both situations, the system is most responsive (and successful) when those with responsibility have appropriate authority. Power cannot be allowed to drift upward to some centralized position.

If the rules and roles are clearly understood, and if power is equal to responsibility at every level, people know who to deal with on any specific question; people also know when they don't have to deal with anyone else, when independent action is appropriate, when it truly is "their call." They have the responsibility, they get the power.

Innate adeptness for change and meeting new challenges and opportunities... It is important to note that rules and roles are often re-cast at moments of coordination or potential change since evolution is not only possible, it is inevitable. Any organization must be willing and able to change to match the environment.

Consider that 50 years ago the statement "*Separate but Equal*" was accepted by the majority of Americans; today this statement is considered somewhere between bizarre and abhorrent.

In the workplace 50 years ago, statements such as, *You are paid to work, not think* were commonplace– and accepted. To any company hoping to survive to see the 21st century such a belief is foolish beyond discussion.

Rights and responsibilities Any parallel drawn between the American quality revolution and the American government would be incomplete if it did not also examine the responsibilities and rights of the citizens. The minimal requirement to be considered a good citizen is an informed vote. Anyone who fails in this respect forfeits all right to complain about any aspect of the government.

Participation... The beauty of the American government system is that any citizen who chooses can become actively involved in influencing policy. Everything from gathering petition signatures to attendance at town hall meetings to running for office are all available to virtually anyone willing to put in the time and effort.

There is in the structure an implied recognition of the potential talent held, sometimes unknowingly, by members of the electorate—and a desire to tap into that knowledge and experience base.

So too with a quality process... Just showing up for work and doing a competent job is a bit like informed voting. It's the minimum; it gives a person a right to voice an opinion. A quality process opens all the other possibilities. This is not to say that a quality process invites anybody to do anything whenever they want. Just as in the political/government structure, there are rules and roles; rules and roles that can be taken advantage of and/or changed if the citizens bring enough discussion and reasoning and work to bear. Just as a voter who takes part in a successful petition drive has shared in appropriate power to make a change in the *way things are done around here,* so too with the member of a team that is the catalyst for a change, large or small, in corporate procedures.

The participation era has already begun At a time in our national history when more and more people who are new to the halls of government—women, non-white citizens of both genders—are becoming players in shaping the future, it is only appropriate that quality processes have provided a path for people other than the CEO and senior executives to make contributions, to exercise appropriate power, and to climb the corporate ladder as well.

Overcoming error can drive quality in all things... During the 1990s, the entire world moved closer to democratic ideals. The monolithic tyrannies are gone or are in decline. While the success of embryonic democracies is not a sure thing, there is cause for optimism. It is good to remember that even the United States didn't get it right the first time. The Articles of Confederation were the governing document immediately after the Revolutionary War and they were anything but a success. By failing to strike a balance between rights and responsibilities, between individuals and community, chaos resulted.

To the eternal credit of our nation's founders, those unsettled conditions were not used as an excuse to return to a less democratic form of government, one with power hoarded at the top. They recognized that the problems lay with the definition of the *rules*, not with the people impacted by those rules.

Those leaders sought a way to still realize their democratic ideal while solving the inherent – and inevitable – problems, and their persistence and success has benefited millions of people in this country and around the globe.

This example can be the source of great hope for companies trying to join the quality revolution. Quality efforts that fail at the first effort are not proof that the basic idea is grievously flawed. Nor are they an indication that the employees of the organization just aren't up to it. They are merely preparation for quality efforts that succeed — provided there is clarity and diversity in rules and roles. What is imperative is leadership at the top. Corporate executives would do well to ignore the latest set of management books and look instead to their history books to learn from the examples of Jefferson, Adams et al. The citizens are ready.

Journal for Quality and Participation, *March 1995.*

AN EXCELLENT ENTERPRISE IF YOU CAN KEEP IT

Benjamin Franklin probably said it best. When a woman stopped him outside the meeting of the Constitutional Convention and asked him what form of government had been decided upon, he replied, "A republic if you can keep it."

In those few words, Ben Franklin encapsulated the political growth of the "soon to be" United States of America: the continuity from the impassioned statement of philosophy embodied in the Declaration of Independence through the loose association under the Articles of Confederation to the recognition that if the nation were to survive, it must be supported by a clear delineation of the relationship between the states and the central power. The leadership chose a republic, but even then, Franklin knew that an alert citizenry would be required to secure its benefits.

Political rights/responsibilities Freedom is a two-sided coin: on the one side, rights; on the other, responsibilities. Most people forget that the bulk of the Constitution spelled out the responsibilities of the respective governmental entities; it was left to the first ten amendments to spell out the Bill of Rights.

We don't talk about the rights/responsibilities duality very much, although we do accept it in our everyday dealings with the government. We insist on our rights. We fulfill many of our responsibilities virtually automatically, even though we may occasionally grumble. There are actions we are compelled to do. We pay our taxes. (*Nasty things happen if we don't.*) There are actions we are constrained from doing. We don't shout "Fire!" in a crowded theater. (*Nasty things happen if we do.*) We comply, and we don't see anything odd about it.

But most of us are uncomfortable talking about responsibility in the same breath as rights. It sounds so, well, pompous.

This duality could help explain much of the debate about 100 percent participation that continues throughout the quality community.

Some rights questions... How, it is asked, can you compel someone to contribute to quality improvement? Isn't it better to make such a contribution voluntary? Doesn't it somehow violate employees' freedom of action to make them serve on a quality team? (*Call this the "rights theory" of quality participation.*)

Some responsibility questions... Then there's another set of questions, an equally unanswerable set, that stakes out the position from the other side. Is there an inalienable right to produce shoddy goods and services? Can an employee refuse to think on the job? How does working with a fellow employee represent an infringement of freedom? (*Call this the "responsibility theory" of quality participation.*)

What is the real issue? The two sets of questions test the extremes. Both beg the main issue. There is nothing inherently wrong in either compelling an action or constraining an action; it's done all the time. Every business requires employees to arrive on time for work; every business prohibits theft. No one gets upset by either.

What is important is recognizing what you are compelling or constraining and why.

The why's of quality... In terms of quality improvement, the why's are fairly clear: customer satisfaction, employee morale, profit and loss, and job preservation are all linked to quality.

The why of 100 percent participation... The basic premise of 100 percent participation is this: the way to achieve quality goals is to provide an opportunity for every member on the payroll to contribute to the fullest.

What is the Challenge?

Konosuke Matsushita thinks that most organizations in the United States don't buy that premise, and he drew some chilling conclusions in a speech to American business managers in 1988:

> "We will win and you will lose. You cannot do anything about it because your failure is an internal disease. Your companies are based on Taylor's principles.[1] Worse, your heads are Taylorized too. You firmly believe that sound management means executives on one side and workers on the other; on one side men who think and on the other side men who can only work. For you, management is the art of smoothly transferring the executives' ideas to the workers' hands.
>
> We have passed the Taylor stage. We are aware that business has become terribly complex. Survival is very uncertain in an environment filled with risk, the unexpected, and competition. Therefore, a company must have the commitment of the minds of all of its employees to survive. For us, management is the entire work force's intellectual commitment at the service of the company... without self-imposed functional or class barriers."

Those Americans who believe in 100 percent participation would argue with his perception. Matsushita has, however, pointed the way to effective worker participation, although there can still be disagreement on the most effective way to achieve it.

Can you compel participation? Actually, yes, participation can become a condition of employment. Some organizations require that everyone on the payroll become a member of a quality team, and that is not in and of itself wrong.

Beyond thinking about it and planning for it... Any organization that moves beyond a declaration of intent, through an exploratory stage of deciding what would improve quality within the organization, and settles on a structure of quality teams can be enlarging the scope of action for every employee.

A quality and participation constitution? Far from shrinking options and violating individual freedom, the organi-

[1] Frederick W. Taylor, author of *The Principles of Scientific Management* (1911).

zation to this point has merely defined a framework for participation.

It may be stretching the analogy a bit, but it isn't entirely far-fetched to compare employment in a firm to citizenship and membership on a quality team to a residency requirement. Even though a team meets, an individual member may not enter into the decision-making process on every issue.

Put in the terms of the analogy, employees may still choose not to vote, but if they do choose to vote, they are aware of how to register their opinions. And everyone agrees to abide by the decisions of the majority whether they vote or not.

Quality Teams: Teams in Deed and Taylorism by Another Name

Not all teams are so benign. Just calling a group of people a quality team does not necessarily make it so.

Teams in name only Ask these few questions to discover how much latitude the team has:

- Is someone not on the team setting the agendas for the teams?
- Do decisions of the team undergo management review amounting to permission to act?
- Are the teams punished for not meeting quotas?
- Are teams only for non-management personnel?

If the answer is yes for each question, then, indeed, there is evidence that they are philosophically incompatible with what most quality professionals envision when they talk about teams. In this case, the format is just a disguise for Taylor revisited.

Teams in deed... Effective quality teams, teams that utilize all of the talents of their members, are virtually self-governing. They identify problems in their own work areas. They present solutions as *fait accompli* for tracking purposes. They are thanked, not hectored.

Most importantly, they are genuinely focused on quality, not management approval. Having teams of managers working on their own issues helps drive this home.

Non-team Alternatives

There are alternatives to quality teams that are clearly less directive; some of them are very effective. Unfortunately, under the guise of protecting employees from "too much pressure" (team membership), many organizations effectively disenfranchise them altogether.

They design elaborate systems, too complex for mere mortals. Or they devise a more passive framework for 100 percent participation, either a suggestion system or calls for volunteers, and then fail to support the structure chosen.

Informed consent and participation... Both suggestion systems and volunteer efforts can allow every individual in the company to participate, and both can be successful. Both, however, require informed consent to really enlarge on the possibilities for individuals and organizations involved.

If every employee in the company is trained to write a successful suggestion, or if every employee in the company has had the opportunity to master problem solving techniques and group dynamics, then, and only then, can these structures provide the same results as trained, standing teams with 100 percent participation. In every scenario, employees are compelled to receive training in how the quality improvement process within the company works.

Ironically, many fierce defenders of the right not to participate in a quality process (except perhaps voluntarily), see nothing wrong with managers making decisions that impact every employee in the organization.

How it can infringe on someone's freedom to provide a concrete, pro-active structure that enables that employee to impact decision-making, and not infringe on that same individual's freedom to tell him or her what to do is a conundrum. At best, such behavior mirrors the philosophical bent of a benevolent dictator; at worst, it becomes the refuge of a petty tyrant.

Pro-active Versus Passive Involvement

Effective 100 percent participation not only provides for every employee to have some way to impact the decision-making process, it also matches authority to responsibility; if you get blamed for something, you have the ability to change it.

What a quality process can do is to create a situation where an employee can exercise this right without jeopardizing his or her job—even when it means bucking the system. That is where a pro-active structure is more effective than a passive one.

Teams in a 100 percent employee involvement structure require every employee to at least look at the issues. When people meet to discuss quality, to solve problems, and to ask questions, there is an effect very like the one Carl Sagan noted when discussing scientific inquiry in a September article in *Parade Magazine*:

> "There is built-in error-correcting machinery. There are no forbidden questions in science, no matters too sensitive or delicate to be probed, no sacred truths. There is an openness to new ideas combined with the most rigorous, skeptical scrutiny of all ideas, a sifting of the wheat from the chaff. Arguments from authority are worthless. It makes no difference how smart, august or beloved you are. You must prove your case in the face of determined, expert criticism. Diversity and debate between contending views are valued."

The title of the Sagan article was "Real Patriots Ask Questions."

There is no real conflict between freedom of action and a structure that enables freedom of action. Just as the Constitution was crafted to preserve the goals of the Declaration of Independence, a company structure can be fashioned to achieve the goals of quality.

Ending at the beginning, it is easy to picture an American business executive in the 1990s walking out of a crucial meeting of the senior management of an organization declaring that, "We have a 100 percent employee involvement quality process if we can keep it."

Journal for Quality and Participation, *December 1991.*

Countries and companies that value individuals attract people. Immigrants continue to stream into the United States over 200 years after the American Revolution; companies, such as Baldrige winner Motorola, have a voluntary turnover rate in the single digits, despite the strain of enormous growth (almost 17 percent or 20,000 employees added in two years).

The parallels between the growth of the American experiment and the quality movement are substantial.

The history of this nation began with a simple request to be treated as full British subjects. What became a full-scale revolution and subsequent separation from England might have ended differently had the boss (king) chosen to treat the inhabitants of the American colonies with the respect they felt they were due as Englishmen. In contrast to King George III, Prussian Baron von Steuben accommodated the American temperament. Training the army, he treated the soldiers as adults, answering their questions, listening to them, and explaining their mission in detail. The previously shabby armed forces emerged from winter encampment a unified force and never lost another battle. Without Baron von Steuben's stereotype-shattering actions at Valley Forge, the revolution might well have failed.

The American Revolution and the subsequent political evolution were unique in world history: the same people who initiated the revolution built a new country and took part in public life long into the peace. There was no internal coup, no purge, no "me-first" power struggle. The intent from the outset was the definition and implementation of a system that would be self-perpetuating and would benefit all individuals.

The remarkable geniuses who defined the United States of America–among them George Washington, Thomas Jefferson, John Adams, Benjamin Franklin, and Alexander Hamilton – were leaders and innovators. They set about an environment in which every person could be heard and in which the accumulated knowledge of individual citizens could be brought to bear for the common good.

Any representative government assigns power and responsibility. When the Founding Fathers spoke of people being *equal*, they meant *equality of opportunity* and *equality of rights*, not equality of talent, intelligence, authority, or responsibility.

LESSONS FROM THE REVOLUTION

Three lessons from the American Revolution relate to the quality revolution.

Leadership

Capital "L" Leadership (what CEOs and other senior executives do when they look out over the horizon and make the major decisions that set the direction of the enterprise) is absolutely vital. It is what Jefferson and others provided and what the President and other officials, both elected and appointed, continue to provide. Small "l" leadership (what is done at every level of the organization day in and day out) is equally important. It is the exercise of informed decision-making at every level of an organization, enabling a government or business to function. In the best of worlds, each layer draws from the information available and makes only those decisions appropriate to their level, allowing those at lower levels to make the decisions appropriate to their levels. Authority equal to responsibility is the goal.

Participation

Every citizen has a right to take part in the government—not as a co-equal in decision-making with the president and other elected officials, but as a source of ideas and as a voter. Participation in government—as candidates, as voters, as volunteers, as workers in community initiatives, and any of the other multitude of options that allow citizens to impact government—is a distinguishing feature of the American experiment. While the level of citizen involvement characteristic of the American body politic is no longer rare among the nations of the world, it was little short of bizarre in the late 18th century.

Unfortunately, everyone eligible to vote does not do so. Nor does every employee submit ideas for quality improvement. That does not abrogate the value of a structure that encourages 100 percent participation. One of the least comfortable features of our republic is that rights come with responsibilities. Where people shirk responsibilities there are consequences; republics and companies can both fail.

Measurement

One meaningful measurement of the American experiment is that the United States has the second oldest constitutionally based government in the world. Day-to-day measurement, however, is essential since measurement is the basis of accountability. Measurement that becomes an end in itself, however, serves no purpose. The constant battles among the three branches of the government are over the choice of measures and over the balance between solely rational measures and more emotional assessments. That same sort of tension can be expected in any quality process. The trick in government or private enterprise is to engage the people being measured (or doing the measuring) in determining how to gather, interpret, and use data.

Perhaps the best slogan for the next phase of the American Quality Revolution is "Do it right the next time!" If an organization has "tried quality" and nothing happened, the fault must be in the implementation. Redefining the parameters, or simply guiding and reacting to inevitable evolution, requires enough self-confidence to admit that change is needed and sufficient wisdom to keep the effort on track.

Executive Excellence, *12th, July 1995.*

2

THE THREE COMPONENTS

Leadership
Participation
Measurement

Long-term success in an organization depends on a blend of activity, growth, and influence in three major categories: leadership, participation, and measurement. Executives ignore any one of the three at their peril.

Without leadership, a quality process lacks focus. It also lacks the example of action. It is not enough for executives to allocate resources and retire from the fray. They must work at actively improving what they do.

Quality is rooted in participation. Without it, a process is reduced to the bleak prospect of impassioned speeches and walls filled with data-heavy charts. The wider the participation, the more potential the process has. Relegating quality to a quality control department limits the impact of quality considerably.

Without measurement, the quality effort is meaningless: How can an organization tell if it is getting better if it does not know where it has been or where it is now? Nor can an organization make consistently smart decisions on what to do next without measurements to guide it. Interestingly, because measurement is the component of quality that is the most rational, this is the one that is most often installed. It is also the one that is most often misused.

What makes the pursuit of quality difficult is the need to pay attention to all three aspects at the same time. In a sense, it is like juggling: You need to keep all the balls in the air and moving at all times. No one pays to watch a person juggle one ball.

3

LEADERSHIP

This chapter gives examples of leadership and points out actions a leader can and should take, particularly in the context of quality. Any attempt to introduce a quality effort to an organization will, of course, require changes throughout the organization—both in personal attitudes and in specific procedures. "Making Change Possible" offers ideas on six specific things a leader can do to help his or her people accept the necessary changes.

"Beyond Charging the Hill and Demanding Excellence" offers "seven aspects of leadership and change" drawn from the study of historic characters. Incidentally, the women mentioned—Molly Pitcher and Clara Barton—are fascinating people. Pitcher fought alongside the men in the American Revolutionary War in the 1770s while Clara Barton founded the American Red Cross during the American Civil War in the 1860s.

"Take It Personally" speaks directly to the senior management team of a quality-seeking company. "Leadership at Every Level" pertains to everyone on the payroll. It offers definitions of terms that are commonly used but rarely defined: authority, responsibility, and accountability.

Finally, "The Secrets of Continuous Quality Improvement" provides concrete suggestions for changes in leadership behavior.

The discussion of leadership is continued in Chapter 9 using the prism of military leadership to look at universal principles.

MAKING CHANGE POSSIBLE

One of Newton's Laws is that bodies in motion tend to stay in motion and that bodies at rest tend to stay at rest.

There is an organizational version of this basic physical truth: Those for whom growth and forward movement are the norm tend to be exemplars of change, while those whose corporate mantra is "this is how we do things around here" tend to make the equivalent of buggy whips until something extraordinary happens. Unfortunately, that something extraordinary is usually extraordinarily painful as well.

Making change possible while maintaining individual and corporate sanity is one of the primary responsibilities of leadership. At any time, leaders are responsible for establishing an environment in which others can reach their full potential while simultaneously completing the job.

In times of transition, maintaining such an environment is especially difficult, but there are specific things a leader can do to make the prospect of change exhilarating rather than frightening, among them are the following:

- Never institute change for change's sake
- Prepare the environment
- Show concern for both projects and people
- Know your people and demand the best from them
- Share knowledge
- Recognize accomplishments.

Never institute change for change's sake If individuals believe that their managers (at any level) are simply following a fad, the bonds between followers and leaders will be strained to, or beyond, the breaking point. Many of the decisions to try re-engineering in the last several years fall into this category.

The upheavals caused by this *let's-try-this-and-see-what-happens* decision-making process not only failed to result in significant economic gains for most organizations, it shattered the trust between senior management and the members of the organizations. This was not wise. As one eight-year-old put it. "Trust is a glass bubble. Did you ever try to fix a glass bubble?"

Prepare the environment Talk about what change means. Is there an alternative? When people realize that change happens regardless of personal preferences, the discussion shifts. The question is no longer *if* there will be a change, but what kind of change will best serve our purposes.

By putting people in charge of change through personal example, through training, and through efforts to pass along all possible information, a leader can build up a bank account of good will with subordinates, so that when major changes occur, subordinates have the skills and willingness to cooperate. Anticipating changes establishes an environment that takes out much of the risk/surprise.

Show concern for both projects and people According to the military, the top two goals of leadership are to:

1. Accomplish the mission.
2. Take care of your people.

Implicitly, the best way to accomplish the mission is to take care of people:

- Act as a buffer to protect subordinates.
- Avoid passing the buck.
- If change puts your people in an impossible position, resist it.
- If necessary, act from the courage of your convictions, even when such a position runs counter to the policy of seniors.

Gen. Oliver P. Smith (US Marine Corps) refused Gen. Douglas A. MacArthur's (US Army) orders to string out his troops in the rush to the North Korean border because he knew that it would squander his troops. He was subsequently vindicated.

Know your people and demand the best from them People live up to expectations. Leaders who have confidence in the ability of their people transfer that confidence. To make sure that confidence is merited, teach and mentor and counsel subordinates. Make sure that individuals have peer support to help them adjust to change. Work as a team. Teamwork plays an important role as subordinates decide how to do things right after leaders have decided what are the right things to do. Demand initiative and accountability. In the words of the *Participation at Donnelly Corporation* text:

> "*One way of thinking about this is to say that not everyone has to be the captain of the ship, but no one is permitted to be an anchor.*"

If someone is acting as an anchor, it can probably be traced back to an autocratic leader who withheld information or rejected ideas. Participative and delegative leaders know their people and their capabilities, both essential in evaluating and instituting change. Keep in mind that people will also live down to expectations. If a manager expects little and shares less, people will eventually give up trying to do their best.

Share knowledge Only if people know where they as individuals and as an organization now stand and where they are headed can they possibly contribute their best effort. An attempt at change without sharing knowledge invites disaster. Make sure that tasks are not only understood, but the *reasons* for the tasks are understood. Keep in mind that what any individual gives to an effort is freely given—and can be withheld. Sharing appropriate information will require the ability to both teach and to communicate. Fortunately, these are skills that can be learned, although acquiring and improving these skills requires an investment of time and effort on the part of leaders.

Recognize accomplishments Like so much in leadership, there is both a rational and an emotional reason to say thank you. Emotionally, people who contribute to an organization deserve to know that their efforts are appreciated; rationally, people who hear an organization say thank you are likely to redouble their efforts.

Change depends on a willingness to experiment... Saying a special thank you for improvements reinforces the belief that the organization values change. And remember to say thank you in a number of different ways. It may be *fair* to present everyone with a plaque, but while half the recipients may put it over the fireplace, the other half are just as likely to put it in the flames.

All of this requires a great deal of self-confidence on the part of the leader Self-confidence, however, is consistent with the demands of leadership. The simple fact is that only the self-confident can lead; insecure people are limited to being managers—until they make the effort to grow past that status into the role of leader.

Change will always be present, but it need not be traumatic The question is one of leadership. When competent leaders are in place at all levels of the organization, both the individuals who make up the whole and the organization as a corporate entity can make change "the way we do things around here." Embracing change as opportunity can make all the difference.

Journal for Quality and Participation, *March 1997.*

BEYOND CHARGING THE HILL AND DEMANDING EXCELLENCE

At the height of the Battle of Monmouth Court House, after months of shared hardships, when the flow of combat could have gone either way, George Washington rode into the middle of the fighting, rebuked the general in charge of the action and personally led his American troops in the fierce fighting.

A big man (at 6′2″, Washington remains one of the tallest presidents of the United States) astride a huge white horse, he made an obvious target. His troops rallied, and an important victory was won.

General Joshua Chamberlain... During the Civil War, at the Battle of Quaker Road, Union General Joshua Chamberlain, a former liberal arts professor from Bowdoin College, Maine, was wounded during the initial charge and lost consciousness. A bullet passed through his horse's neck, hit him in the chest, and traveled around his rib cage and out his back. When he revived, he saw that his troops had begun to retreat. Chamberlain, bleeding front and back, rode his horse (also bleeding profusely) back into the fray and led his troops into the now-oncoming Rebel forces. The result was a Union victory.

Both men engaged in acts of inspirational leadership that changed the outcome of important battles in an instant. And both stories explain why inspirational leader- ship is often associated with bold action and moments of crisis. Few of today's corporate leaders have such blood-stirring opportunities.

Other aspects of leadership... There are, however, other stories about both men that are equally telling. Most people have heard of Washington's farewell to his troops in 1781, where he bade them preserve the new nation they had created together.

Fewer have heard of Chamberlain's generous spirit in calling his troops to attention to honor General Lee at Appomattox. Bold action had become quiet reflection; crisis had become change. Yet, both these events were also acts of inspirational leadership.

Seven aspects of leadership and inspiring change Much can be learned by modern-day business people from the actions of people such as Washington and Chamberlain. Look again at the above examples for the things they have in common:

1. Both men were technically competent. Washington's accomplishments are well known. Chamberlain by dint of extensive self-education efforts became one of the most knowledgeable leaders of the Union Army.
2. Both believed passionately in the goal they were trying to achieve.
3. Both cared about people.
4. Neither asked people to do what he would not do himself.
5. Both trusted others to follow their example.
6. Both were masters of the gesture, large and small.
7. Neither would tolerate incompetence.

By fulfilling these same criteria, Molly Pitcher and Clara Barton became equally well-known figures of those same wars.

How would these criteria play out in a corporate setting? What follows is a description of the actions of one paticular leader during the first few years of the implementation of a first-rate quality process.

Portrait of a Complete Leader

Aubrey K. Reid was the company president of the Paul Revere Insurance Group. He had been chosen for the role several years earlier because of his personal success in the sales department of the company, his extensive knowledge of the insurance industry in general and the company's products in particular, and his reputation as the smart, hard-nosed business manager.

In May 1983, the senior management team of the Paul Revere Insurance Group decided to initiate a quality process as a means of regaining market share. After decades as the market share leader in their particular market niche, they had slipped to second place. Their pride had been stung; they wanted to be first again.

Effectiveness, humility and trust Since becoming president, while burnishing his reputation as an effective, bottom-line manager, Aubrey had put the company on a first-name basis and had tried to make life a bit less regimented for the employees. He adopted a delegative leadership style to help change the company from a top-down, patriarchal one to a bottom-up, open model. In line with this, he chose to stay off the quality steering committee and trusted his staff to make the formative decisions about quality.

The *Quality Has Value* process was set to be launched on Friday the thirteenth, January 1984. At the kick-off ceremony, Aubrey contributed a few words of greeting and support and then sat in the audience with the rest of the team leaders while members of the *Quality Steering Committee* ran the show.

Personal involvement in measurement and recognizing excellence At Paul Revere, every employee was on a quality team from the outset and, in the first year, quality team leaders were the natural work unit leaders. In December of 1983, Aubrey attended the firm's *Quality Team Leader training*. Aubrey led the quality team known as *The Big Guys*, consisting of his direct reports, all vice presidents and senior vice presidents. If any employee of Paul Revere wanted to check on how serious the president of the company was about quality, he or she could call up the *Quality Team Tracking Program* on the nearest terminal and look at *The Big Guy's record.*

During the first few months of 1984, Aubrey frequently conducted quality team award ceremonies. There were lots of these as the first level of recognition could be reached either through the implementation of ten quality ideas or the saving of $10,000 (annualized savings—hard dollars or soft). A couple of times a week, employees would see Aubrey headed somewhere in the building, a collection of *Quality Has Value* pins and other items under his arm.

Once the process appeared to have found its feet, Aubrey's personal involvement in team award ceremonies was reduced somewhat, and the two co-chairmen of the *Quality Steering Committee* conducted the majority of the ceremonies. By mid-way through the second year, Aubrey normally only conducted the ceremonies that marked the second time a team reached the third level (*Double Gold*) of team recognition. But, by then, one or more teams reached *Double Gold* every other week or so. To further encourage the change in the corporate culture, Aubrey oversaw two programs: *PEET* and *Quality Coins. PEET* stood for *Program for Ensuring that Everybody is Thanked.*

PEET (Program for Ensuring that Everybody is Thanked) and Quality Coins...

PEET On the first of each month, every member of the senior manager committee, 25 senior managers, Aubrey, his direct reports, and many of their direct reports received a *PEET Sheet* with the names of two quality team leaders. Each senior manager was expected to visit each of the two team leaders noted on the *PEET Sheet* (on the team leader's turf) sometime during the month.

Quality coins The *Quality Coins* were instruments for instant recognition. The senior managers were urged to make an effort to "catch someone doing something right" and say thank you. The coins could, at the option of the recipient, be cashed in for a free meal in the company cafeteria.

The leader leads through personal consistence and persistence... The programs, launched several months apart, both started big and quickly slumped.

Intellectually, Paul Revere's senior managers acknowledged the need for change and warmly received the announcement of the two programs for breaking down old habits; emotionally, they felt breaking habits had little priority. But Aubrey made his *PEET* visits. And Aubrey found time to identify individuals who had stretched a bit to do something extra and gave them *Quality Coins.*

To assist the senior managers in their resolution to do what they had all agreed was a good idea, Aubrey announced the implementation of a new measure since, after all, you can't manage what you don't measure. After explaining his decision to the senior managers at the monthly meeting of the 25 of them, he required the director of the quality process to provide him with two lists at the first of each month: the *PEET Sheet Report* and the *Quality Coin Report* (with a monthly goal of five coins).

Responsibility and accountability... The two reports were the last two items on the meeting agenda every month. (The progress of the *Quality Has Value* process was always the first agenda item.) If a senior manager had not found the time to visit with the two designated team leaders in the course of an entire month, or had not been able to identify at least five people who had done something right in their department in the course of that month, he or she had the opportunity to explain his or her difficulties to Aubrey and all of his or her peers.

Mixing leadership styles... The Quality Has Value process gave Aubrey an opportunity to demonstrate that his leadership skills extended far beyond his ability to make the right decisions about the composition of a new insurance policy or to sell that policy once it was available.

In return, by being a leader, and not just a manager, he gave the process much of its soul. By his careful mixture of authoritarian, participative, and delegative leadership and his own example as a participant at all levels. Aubrey inspired the culture shift he desired so fervently.

A moment's reflection on inspirational leadership...

Many employees' fondest memory of Aubrey K. Reid comes from the *Quality Celebration* at the end of that marvelous inaugural year of the *Quality Has Value* process. Every team, save one team of middle and senior managers, had reached the first (*Bronze*) level of recognition and over half of the 125 teams had reached the third (*Gold*) level. Amid balloons, banners and music, Aubrey took part in handing out year-end awards to many individuals and several teams.

The wrap-up of the *Quality Celebration* was to be a short speech by Aubrey. He began by speaking in general terms about the company and the impact of the quality process. He took just a moment for a gentle jab at the one team that hadn't made it to *Bronze*, noting that it was a team composed of fairly senior folks. Then he began to talk about the people of Paul Revere.

By the time he said, "I am just so very proud of all of you," his voice cracked. He stepped back from the lectern to compose himself. During the 30 seconds or so that passed before he could resume his remarks, there was absolute silence. After he concluded his remarks, he received a long and affectionate standing ovation.

Aubrey retired just a few months after Paul Revere received a site visit for the first Baldrige award. He turned Paul Revere over to the senior vice-president for operations, who had himself been one of the two co-chairman of the *Quality Steering Committee*, on January 1, 1989.

Leadership without Bombs Bursting in Air

Unlike Washington and Chamberlain, Aubrey didn't ask for life-threatening acts of heroism, but he did require career-changing acts that demonstrated a commitment and willingness to follow his lead, perhaps into brand new experiences, both intellectual and emotional. He asked others to become better leaders themselves.

It's not just in war or just at the top... Acts of inspirational leadership don't come just from war heroes and CEOs. They can be found at every level in an organization. Consider the following story about a Federal Express employee, related by its chairman, Fred Smith, at the *National Quality Forum VI* in October 1990:

> "The building rumbled and shook to the tremblers (sic) of the October 1989 San Francisco earthquake, but Federal Express courier Maurice Jane't continued to scan each package he was picking up at our company. He struggled to get them down nine flights of stairs through rubble to his waiting van and on to the airport just in time for the plane. An extraordinary example of personal commitment," wrote CEO Gary Moore (of Hatachi Data Systems).

Competence, passion, caring, master of the gesture Maurice Jane't's actions had much in common with those of Washington, Chamberlain and Reid. They were acts of inspirational leadership.

Journal for Quality and Participation, *March 1994.*

TAKE IT PERSONALLY

The central lessons of the Eastern-Lorenzo-Ueberroth pageant are first that virtually everyone on the payroll has a strong opinion about the top executives in a company, and second, that if opinion is negative or inaccurate, it is difficult to ameliorate. For better or for worse, an executive leaves tracks wherever he or she goes.

At no time are those tracks more carefully studied—or more strongly criticized—than during a period of proposed or on-going change. When an executive begins calling for "change in the corporate culture" or "change in the way we do things around here," the first question silently asked is, "And how are *you* going to change what *you* do?"

If the response is, in effect, "I'm going to change the way you do things" or, worse yet, "I'm going to hire some people to tell you what to change," the project is doomed—regardless of how expensive the consultants, how elaborate the program, or how lavish the roll-out.

AN EXECUTIVE OBSESSION

This is especially true if the change is aimed at improving quality. Quality improvement requires the personal, constant and consistent involvement of top executives. In fact, the word "commitment" may be too weak; "obsession" may be more accurate.

A willingness to sign checks (usually large) for consultants and to make an annual speech (usually short) does not constitute commitment, much less obsession. Talking with employees at all levels, casually and formally, about quality; funding the classes and the structure needed to sustain a viable quality process; and being personally involved in the day-to-day operation of a unique quality process–that is what is needed.

There are at least three primary reasons that leaders of an organization might choose to begin something that will call for them to critically assess how they do their own jobs, to transfer considerable power down the chain, and to radically realign their own priorities: money, money and more money. Quality makes money.

Secondary reasons include the preservation of jobs, the creation of jobs, the positive impact on the morale of all employees, and the simple fact that a quality process is not only wonderfully satisfying, it can be a great deal of fun.

All of these benefits are readily available so long as top management focuses tightly on quality—not on productivity and not on cost-cutting (these are by-products of a well-run quality process).

THE QUALITY PROCESS

Executives must take responsibility for the definition, implementation, and continued activity of the quality process. It begins with personal development and learning. Having a group of executives attend an expensive school, while useful and perhaps even necessary to ensure a certain knowledge level, is not enough. In the eyes of people who rarely go on a company-sponsored trip and whose favorite in-house training class was cancelled last month for lack of funds, attendance at such schools can be viewed as recreation. Such attendance has the added disadvantage of being a "one shot" occasion, when what is called for is developing new habits.

That is why a series of in-house meetings featuring discussions of books and articles on quality has greater impact. Use this issue of *Executive Excellence* or the March issue of the *Journal of the Association for Quality and Participation*. Talk with some of the experts in the field. Send different executives to courses taught by experienced trainers/presenters whose books or articles are studied in the sessions. Ask each executive to teach his staff what he learned to speed the group learning process.

The object is to reach consensus at the top on whether a quality effort is appropriate or not. The boss cannot make the decision in isolation and storm ahead; if the top managment team hasn't decided that this is where they want the company to go, the results will never approach full potential.

Once a decision to proceed is made, the top management team should establish a Quality Steering Committee (composed of top executives from each division or major department). The committee's task is straightforward: to define a quality process that will formally enroll every member in the effort to continually improve the organization. Funding must be made available; money well spent on a quality process is money wisely invested.

Meetings of the Steering Committee must have high priority. The definition of "tomorrow's company" must have precedence over all but the most calamitous problems of today. For reasons of style and substance, hold the meetings in the most noticeable conference room with religious regularity. Attend yourself. If employees are hearing (at least through the grapevine) all about this "new quality process" in the company's future but then see that the executives aren't making it to the Steering Committee meetings, they will excuse themselves immediately.

Another valuable action is regular discussion of the quality process in every top management meeting. Ideally, it will be the first item on the agenda (so that it doesn't fall off the end of the schedule in a rush to finish). Once the habit is set, the executives can carry the procedure to their own staff meetings. At this point, it might be time to notify the company newspaper to begin publicizing the process company-wide.

Company executives (not outside consultants) should decide what elements make up a company's quality process.

Certain concepts are common to virtually all programs, systems or approaches. These need to be adopted. The specific techniques appropriate to implement these concepts must be adapted to the particular corporate culture at hand.

One key decision to make up front: who will be formally enrolled in the process when it is finally launched? The only logical answer, and the one that is most difficult to implement, is "everybody." Any lesser response invites the question: "Why are you leaving them out?"

"Everybody" includes top executives. Whatever rules the company sets for participation in improving quality, they must apply across the board. When it comes to quality, everybody on the payroll has equal responsibility. If employees are required to attend classes on quality or to be members of a quality team, that applies to every employee from the CEO to the newest hire.

While the Steering Committee is progressing with its work, all of the company's executives can lay the groundwork for the new process. They can test some concepts, and talk about them—all the time.

They may also start "listening down." Normally, American business people are taught to listen up and proclaim down, reflecting the assumption that superior knowledge is always higher up the corporate ladder.

Listening down requires getting out and asking questions of real people, to include a large number of that most maligned of working classes, middle managers. It is precisely this kind of personal involvement that will insure that a quality process will retain its vigor, and its financial benefit to the company.

Talk about quality at every opportunity. Take part in training classes, either as a regular speaker or as a drop-in. This invites people to ask questions and make comments about the quality process with support for the company's training program.

Begin thanking employees throughout the company through a program of recognition, gratitude, and celebration, and involve top management in every award presentation or congratulatory celebration, big or small. These award ceremonies, often no more than 10 minutes long, afford you the chance to ask successful employees what they have been doing. The information gained will be clearer, and more useful, than most of what remains in official company memos after they have been carefully worded and even more carefully screened. Regular, and frequent, trips throughout the company to say "thank you" also increases your visibility and underscores your commitment to creating and maintaining an environment in which continual improvement is the norm.

Informed, active commitment to quality is the responsibility of top management. It will be even more critical in the future as an ever-increasing number of competitors challenge companies in both manufacturing and service industries. Only organizations whose leadership is personally and obviously involved in the quality effort will have the flexibility and drive necessary to take full advantage of the talents of every person on the payroll.

Executive Excellence, *May 1989.*

Every leader needs to understand three words: *authority, responsibility,* and *accountability.* These concepts are pivotal to a successful quality process.

- *Authority is the legitimate power of leaders to direct those subordinate to them, or to take action themselves within the scope of their positions.* If a company wishes to benefit from the impact of leadership at every level, a qualifier is necessary: part of a leader's authority is delegated to subordinate leaders who in turn delegate authority down the chain of command until each employee has sufficient authority to carry out his or her tasks.

  What are the guidelines for this delegation of authority? When a person is given responsibility for a mission, he or she should also have sufficient authority to carry it out. Simply put, if a person's corporate future (and that includes something as basic as the possibility of earning a raise) is on the line, he or she should be involved in the decision-making.
- *Responsibility is a requirement to act, an obligation to use the authority that has been given to the employee.* For leaders at every level, authority and responsibility must go hand-in-hand. A key element of leadership training will be preparing people to accept responsibility. After many years of being denied either responsibility or authority, some employees have grown comfortable in a passive role.
- *Accountability is the reckoning, wherein leaders answer for their actions and accept the consequences, good or bad.* As such, it is a cornerstone of leadership. It is the final act in the building of the leader's credibility.

The relationship among authority, responsibility, and accountability is not complex. When employees are granted authority, they are obligated to use that authority to accomplish what needs to be done. Employees are responsible simply by virtue of having authority. After carrying out their responsibilities by using their authority, they become accountable for the results.

You're more likely to have leadership at every level when these fundamental concepts mean the same to all employees. The chances for confusion as to exactly what is meant when someone is told, "You are responsible for this" are dramatically lowered.

Leadership at every level begins with the willingness at the top to delegate authority down the line to the appropriate levels, and the preparation of the men and women at every level to seize that authority and its concomitant responsibility and accountability. In organizations where executives still micro-manage the activity of employees, this may mean a wrenching change. Trust is one of those emotional aspects of leadership that takes time to learn.

As with quality, the creation of an organization in which leadership is understood and practiced at every level is both a sound economic investment and a never-ending challenge. It is the growing awareness of the inextricable relationship between quality and leadership that is now positioning America for what Joseph Juran predicts will be a Century of Quality. The relationship is so intimate that it is fair to say that if leadership and quality are not exactly flip sides of the same coin, they were at least struck in the same mint.

Executive Excellence, *Dec. 1994.*

THE SECRETS OF CONTINUOUS QUALITY IMPROVEMENT

Trust workers enough to give them the authority they need to meet their responsibilities. Support them with tools, training and work processes that enable them to excel and improve. Top management should make clear to employees...

- What's expected of them—not merely the tasks outlined in their job descriptions, but a commitment to do whatever is in their power to satisfy a customer, whether it's explicit in the job description or not.
- What the stakes are for both the individual and the company.
- Get rid of the old baggage of "scientific management." Frederick Taylor, father of scientific management, said: "Any change the workers make to the plan is fatal to success." That is the *false* basis on which US managers have worked for generations—the belief that management is the art of routinely and systematically transferring the ideas of managers and the skills of a few technocrats to the hands of workers down the line.

To survive in this complex and fast-changing world, businesses need the commitment and ideas of *every* employee.

Essential now to sound leadership... Recognizing the need to change to a new relationship between management and non-management.

- Make clarity of purpose—and follow-up of the way improvements will be implemented—a priority. Improvements cannot be generated simply by the chief executive officer proclaiming: "This is now a quality company... and each of you is now empowered to work smarter. Go out there and do it!"

Survey after survey reveals a wide gulf between executives' assessments of how well things are going and the assessments of workers. Executives typically claim they have given workers explicit authority to suggest improvement. But workers report that they rarely, if ever, are asked to participate in making significant decisions about their jobs.

Important: Don't rely on voluntary suggestion systems to which employees can contribute or not contribute, that respond slowly or not at all to suggestions, or that fail to encourage workers to implement their own ideas.

Unless executives and employees are clear about how to improve quality, there will be great waste of effort—or not effort at all.

Lead rather than manipulate You cannot manipulate people into doing high-quality work. Incentives are manipulative—and they are not the best way to get quality efforts from people. Incentives encourage only a small part of the effort most employees are willing to give to make the company successful.

Gratitude and recognition are more powerful than merely writing checks. In one organization where managers give out specially minted coins as an immediate expression of thanks for a job well done, only 25% of the coins are ever turned in for the $10 cash reward. Employees hold on to the coins to remember the moment.

The key to genuine motivation is sincerity—in caring about the welfare of everyone in the group. Without the ability to care, it's impossible to be a great, or even a good, leader.

Employees who perceive that a leader doesn't care about at least the basics—a clean and comfortable workplace, the right equipment and resources, parking, safety—consider the person a *poor* leader.

Make proper use of the variety of leadership techniques available Be confident that you can learn them—and when to use them. This means sometimes encouraging the group through participative management and sometimes imposing decisions from above. *Authoritarian leadership has its place when*:

- The leader is the only one with all the necessary information to accomplish an assignment.
- A decision must be made under severe time pressure.
- Subordinates have high morale.

But authoritarian leadership backfires if it's used all the time. Participative and delegative styles of leadership can produce higher levels of morale—creating an environment in which employees respond eagerly to a sharp order because they realize that this is an unusual situation and explanations will follow when time permits.

Leading in a *participative* way involves bringing people into the discussion of a situation, helping them understand priorities, encouraging them to come up with options and alternatives and to reach consensus. But the leader makes the call and explains the reasons for the decision—based on the input of the group.

Delegative leadership is inextricably linked with trust. The leader allows employees to make decisions about how a job is done while remaining responsible for the results. There are two advantages—development of employees and time for the leader to do other things.

Help middle managers survive the shift to quality management that puts more power into the hands of those they supervise As companies flatten their structures and give more authority to teams and workers on the front line, middle managers must change from giving orders to being coaches and resources.

Not all of them can make the shift. Some will have to be let go. But leaders must help managers make the change, if they are flexible enough to do so. Training in leadership and quality techniques supplemented by information on personality and decision-making styles is a good beginning. The best middle managers have volumes of useful information in their heads about the company, its processes, its customers and its suppliers. They can often stop mistakes from being made during a time of turbulent change.

Expect positive results quickly, but don't be disappointed if they're not always super-dramatic Given an opportunity, employees are eager to suggest ways to eliminate long-standing problems and to improve the way work is done.

People are ready to think. They will be eager to use what they learn quickly. *Key:* Encourage them. From practice and from their own contributions, you'll find out if they need more training. Consultants should be there to work *with* you—not to do things *for* you or *to* you.

Be prepared to pay for significant change You'll never get things done by just talking about the need to change. Besides money, you'll have to invest time, ego and effort to change the company so it constantly strives for improvement and higher quality.

Employees are all well aware of how you spend your time. If you expect them to take change seriously for themselves, they are going to want to see changes in management's behavior, too.

Get out of your office. Thank people personally. Congratulating people on their accomplishments should be noticeable. If people are being trained in new techniques and on new equipment, take time to learn something about it yourself. Ask people what else they need. Ask them what they think needs attention in the company.

Board Room Reports, *Dec. 1, 1992.*

4

PARTICIPATION

The reader should be warned: This book's emphasis on 100% employee involvement is not the norm among "quality authors" or "quality consultants." In fact, organizations that fully involve every person on their payroll in their effort to continuously improve are rare. This is puzzling when there are no sound reasons, either financially or philosophically, for an organization *not* to take advantage of the knowledge and experience of everyone to whom they are giving a pay check. In the popular phrase Total Quality Management, total should mean total. Total should mean everybody. And yet, in the majority of cases it doesn't.

"Participation: Starting with the Right Question" presents the argument for 100% employee involvement by comparing it with the other options. "Quality—Down to the Roots" describes various ways in which an organization might actually go about accomplishing that goal. "The 100% Solution or Greater Non-comformity" looks at an even richer opportunity: involving 100% of each employee, that is, taking fuller advantage of each employee's wide range of talents. In virtually every organization in the world, people use only a small percentage of their personal talent on the job. Organizations don't give them the opportunity to do more.

PARTICIPATION: STARTING WITH THE RIGHT QUESTION

Three themes that run through virtually all quality processes are identified: leadership, participation, and measurement. In this we will concentrate on the second of these, and it is there that the right question is required, particularly if the goal is "total quality management." The phrase "total quality" cannot be used with any integrity unless every single person on the payroll is formally, actively, enrolled in the effort to continually improve the organization. Any level of participation less than that should be called PQM—for Partial Quality Management.

Quality is often described as "a journey, not a destination," and Mao Tse-Tung was right in saying that "A journey of a thousand miles begins with but a single step." Often left unsaid, however, is the simple fact that if that first step points the traveler down the wrong path, it is going to be a frustrating journey.

In our new book *Quality in Action*, we present the history of quality within a framework of several questions. Beginning with the right question, like taking a step in the right direction, influences all that follows. Be aware that any question that tries to identify a subset of employees whose primary responsibility is quality while the rest of the employees continue to do "real work" is a step in the wrong direction.

Chronologically, the first question that determined participation trends was "To whom should we assign this quality thing?" or a variation along the lines of "Who can we make responsible for quality?" Up until a decade or so ago, this self-limiting question was practically the only question asked. It could be answered by management picking a small number of specific individuals whose role would be to verify the "quality" of the work of the majority of the employees.

In the case of manufacturing operations, these people (normally under the title "Quality Control") were the modern day equivalent of Horatio-at-the-Bridge. Instead of standing on a narrow span over a river, however, they stood at the door to the loading dock, the company's last hope of stopping "bad stuff" before it was piled onto trucks and distributed across the land.

This approach was—and remains in many companies—a game: a game played with real dollars and real jobs. "Think this is OK?" "Ahhh, let it go... let's see if the quality control guys catch it." Even service got caught up in the game. The service industry equivalents of this approach tended to stand next to the door to the mail room or listen in on telephone conversations. Their efforts were even less effective than those of their manufacturing peers.

Historically, a more expansive version of the question followed. While still self-limiting, it paved the way for a more open approach to participation. This question, "Who can we get to volunteer to do this quality stuff?" was often the beginning point of a quality circles operation. Quality circles nearly always began with a bang and, almost as frequently, ended with a whimper as they self-imploded due to lack of lasting support.

What lack of support? Consider quality circles in most organizations. After volunteers had identified themselves, management began constructing roadblocks. First, membership on quality circles was almost always limited to non-management (under the theory that management did not need improvement?). Then, the circles were assigned projects (after all, these were only non-management personnel and, thus, not skilled enough to pick out areas for improvement on their own). To insure that the group did not move too quickly, each of the circles was assigned a facilitator ("Ah, ah... you've skipped step 4!"). Finally, when a circle did have a solution worked out, they were barred from implementation until they made a presentation to management (when they could gather a quorum of executives, that is) and received a written approval (usually promised "within two weeks"...). It was a frustrating business at best.

Despite all the handicaps, quality circles did have—and continue to have—some extraordinary, albeit isolated, successes. What quality circles proved to American management was that, given half a chance, the American worker is capable of marvellous feats–because that is what quality circles gave the American worker—half a chance.

Oddly enough, the version of the "who should we include" question most often used today (and, not coincidentally, most often recommended/sold by leading consultant firms) does include management. It is, in fact, a type of high-level quality circle. With this approach, several teams are formed consisting only of senior staff. These chosen ones are packed off to a week or more of expensive schooling, and they then proceed to make decisions about how various operations in their subordinate departments should take place. These new procedures are announced, in the name of "quality," to the people who actually have to make the changes/improvements. And then the executives go back to business as usual.

This methodology is very neat, very tidy, and keeps all power exactly where it was before the top management team ever heard of quality—right at the top. While it is appropriate for managers to make decisions that affect subordinates, that is not all there is to quality. A quick litmus test of a quality process is to look to see whose behavior is being forced to change as a result of decisions being made under the quality banner. If it is usually somebody other than the decision maker(s), there is room to question the sincerity of the participants.

There is a sense that quality should no longer be voluntary, no longer be the province of the few, but there is little agreement about how to correct these shortcomings. Perhaps the most honest way of assuring participation (and this has the advantage of being so straightforward as to not require an expensive consultant's help) would be to give each member of the top management team a printout listing every person on the payroll and telling him or her to highlight the names of all those people whom he or she thinks should be included in the effort of improving quality. When consensus is reached on a list of names, the process can be launched.

Honest and quick though it may be, it is still self-limiting. Back up several sentences. Picture again the top management and their lists. Gather them into one room, ask them to ready their highlighter pens, and pose a different question: "Who can we afford to leave out?" If the subtleties of the question seem to evade anyone, the question can be rephrased thus: "Please highlight the names of all those people on the payroll who you think incapable of getting any better."

Management can leave anyone whose name is highlighted out of a quality process, on one condition. The president or CEO must be willing to walk up to the to-be-excluded people individually, look each of them straight in the eye and say, "We've had a meeting of top management and we've decided that you are as dumb as old dirt. In fact, we took a vote and unanimously agreed that none of us think you will ever have another original thought the rest of your life. So we're leaving you out of the quality effort." Everybody else is in.

That is the only philosophy that truly underpins a process that dares to call itself "Total Quality." Unfortunately, the day-to-day operation of 100% participation poses challenges—not all of them mechanical.

First of all, participation of every person on the payroll is not possible unless the top management of the organization has accepted the idea that the personnel department has been hiring adults. Additionally, it must be believed that the overwhelming majority of these adults are willing, if not anxious, to be a part of the future success of the company.

This is not a call for some large idealistic quasi-democracy in which "town meetings" are held to decide important matters. This is a business process that allows everyone to contribute at their level. If there is someone who consistently abuses this power, who maliciously twists this opportunity to participate in building the future organization, fire the bum.

As with most aspects of quality, there is no one-size-fits-all set of procedures available. What is important is that the procedures defined are understandable, accessible, and underscored by the philosophy that everyone is "in" from the beginning.

Success will hinge in part on how the 100% employee involvement quality process is presented. If the workforce is suddenly told that contribution to the effort to continuously improve is now mandatory, followed by a schedule (or deadline) for creativity, the results will most likely be disastrous. Skip debate on the topic entirely. If everyone has the same responsibility and the same authority *vis a vis* quality, if everyone contributes to a suggestion system or participates on a team, then quality improvement is not extraordinary. Stated and implied, the attitude that works is, "Part of what we do around here is get better all the time. You're here. We look forward to your being part of this effort."

One of the more straightforward ways to implement a total quality effort is to break the entire company up into teams of 8–12 people. These quality teams then meet on a regular basis to decide how to do better whatever it is they do. Unfortunately, attempts to follow this line of action have occasionally led companies into ruinously expensive quagmires. If not bogged down in advance by excessive training, they are brought to a halt by overly elaborate follow-up procedures and overheads.

Simplify the training. It is often recommended that, before any progress can be made, everybody needs to receive 3–5 days of training... none of which is free. Aside from building the pension funds of various consultants, much of this training has only marginal value. Imagine for a minute that there are 10 employees seated around a table, prepared to talk about improving their own procedures. Imagine further that they have all had 2–3 hours of orientation training about this new quality process—delivered by a senior executive within the last two months.

How many of these ten employees (or executives—remember: 100% means 100%) need to know how to run a meeting in a participative manner and how to solve problems? One. The team leader. There is no practical need for ten experts on the Pareto Analysis before any activity can begin; there is no financially justifiable reason to delay the initiation of improvement until everyone understands all about the various theories that underline quality efforts. Since quality processes are meant to be permanent, there is plenty of time to train the other 90% of the workforce—financed by savings from quality improvements. In short, a 100% quality effort can be launched after having provided detailed training to approximately 10% of the employees. The other employees will learn from their team leader during the year and in formal classes when it is their turn to be a team leader.

The other end of the quagmire referred to above is the building of elaborate structures to run the quality effort. For a company of 2500–3000 people, there should be no more than 4–5 people (outside of trainers) whose job is to nurture, assist, guide, maintain, and track the quality process. They are a clearing house for quality—not solely responsible for quality. Any more than that will inevitably lead to meddling. The insistence on the recognition of the employees as functioning, contributing, thinking adults applies to the quality professionals as well as to top management.

Participation in the effort to continually improve an organization comes close to being a "worker's right." To truly respect the dignity of the individual, that person's ability to contribute must not only be recognized, but also made possible. To only do so for a limited number of individuals, even if a stated goal is to "achieve 100% participation" within a stated number of years, is to put everyone else's dignity "on hold," an indefensible position.

The Quality Observer, *Sept. 1992*

QUALITY—DOWN TO THE ROOTS

The best looking lawn in the neighborhood requires work. How much depends in large part on whether the objective is to win a contest being held in one month (or to prepare the house for a quick sale) or to establish a lawn that will look good year after year.

If a cosmetic fix is all that is required, forget soil preparation: get rid of the obviously ugly parts, put down a lot of seed, spread fertilizer, and water heavily. Instant results–but the timing better be right so that the contest is won or the sale is made before it is evident that the beautiful ground cover has inadequate roots.

If, however, the objective is long-term growth and results that just keep improving year after year, a great deal more is needed. Beginning with the soil, the homeowner must analyze what adjustments have to be made to provide a hospitable environment. Deep roots ensure that initial results survive any test.

PLANTING QUALITY ROOTS

The same guidelines apply to improving an organization. Top management must first decide whether their quality objective is cosmetic or deeply rooted. A deeply rooted commitment to quality encourages ongoing active employee participation, shares authority, and rewards small ideas. There are other options. The layers of commitment available are, in succession: Suggestion systems, quality circles, task forces/action groups, 100 percent employee involvement, and self-managing teams.

Suggestion systems The first, suggestion systems, allows top management to retain all control while benefiting to some degree from the experiences and input of their employees. There have been some remarkable success stories with this approach, despite the pitifully small percentage of employees who actually ever submit a suggestion.

Comfortable, but... This is a comfortable procedure for management in that they can schedule everything—from reviewing the latest batch of suggestions to implementing ideas deemed worth—at their convenience. They can even change the ground rules with no fear of having to explain themselves to anyone. Suggestion systems encourage large ideas; who is going to bother to write down a small idea and wait for approval? Most recognition is in the form of a percentage of dollars saved, another disincentive for small ideas.

Quality circles The next layer of top management commitment and employee commitment and employee involvement is quality circles. Brought to this country in the late 1970s, these were, and are, a clumsy adaptation of what it was thought was going on in Japan at the time. The voluntary aspect of the circles–a major tenet of the movement—was based on a now-admitted misreading of a book by Dr. Ishikawa.

More responsibility but no more authority... Quality circles were a significant step in the right direction. At least the idea of taking advantage of the problem-solving power of a group of concerned individuals was brought to bear. As quality circles were played out in most organizations, however, management maintained a high degree of control and only a small percentage of the workforce was involved. Oftentimes, circles were assigned big problems to study, and almost always, they had to make a presentation to a management committee before they could proceed with implementation. Management itself was rarely on a circle, but retained the authority to make changes while sharing only the responsibility for a solution.

Task forces and action groups The involvement of management is one of the trademarks of the next layer of quality commitment and involvement, task forces or action groups or quality committees. Declaring that their organization was "committed to total quality," a group of managers would either go off to a course together or bring in a certified guru to instruct them in the ways of quality. Having been sold on the benefits of the concepts of quality, they would then form a series of task forces to work on problems defined by top management as ripe for solution. These tended, not surprisingly, to be big dollar items.

We'll fix it for you... These groups were usually made up of men and women from all levels of management with, perhaps, one or two non-management personnel. Their solution would then be announced with appropriate fanfare, and employees would be told about the new ways to do things in the name of quality. After solving their assigned problem, the task force would disband and another would be formed to attack a different problem.

The litmus test to judge any group's true commitment to continuous improvement is to take note of whose behavior is affected as a result of its decisions. If the group itself has to make changes, commitment can be demonstrated by subsequent behavior. If the only required changes are in someone else's work habits, commitment is open to question. Task forces, more often than not, fail this test.

Task forces to produce improvements, and the positive impact on the company as a whole are addictive enough to keep the process alive for quite some time, but there is one major flaw. This approach can be brought to a standstill simply by not appointing any more task forces. Just as an indifferent gardener can kill a lawn with shallow roots by forgetting to water, a quality process based solely on task forces can be killed by neglect.

One hundred percent involvement A true 100 percent employee involvement team effort is far more difficult to kill. At the third anniversary of one of the pioneer efforts in this country (the Paul Revere Insurance Group "Quality Has Value" process, now in its seventh year), one of the middle managers at the company observed, "You know, if top management were to decide tomorrow that we'd done enough of this 'quality stuff'and cancelled the process, it would take about two years for it to die. People are just too used to having control over what they do."

Sustainable growth... By simultaneously addressing the questions, "Are we doing the right thing?" and "Are we doing things right?" a well-structured 100 percent employee involvement process can drive the roots of quality almost irretrievably deep. In this approach, every single person on the payroll is enrolled and expected to participate on a quality team; everyone is trained to contribute to "doing things right." In addition to "doing things right," management also has the task of determining "right things" to their subordinates.

Such teams have several advantages:

- They involve everyone
- They are ongoing
- They set their own agenda for improvements
- They have authority matched to their responsibility (if the team is blamed for the results, they are authorized to make changes which assure the desired outcome)

With a recognition system that rewards teams for both large and small ideas, they are able to capture small improvements. They have one other attribute: there's no going back to a traditional "all direction and decisions from the top" approach without disastrous consequences in terms of morale and productivity.

Self-managing teams Even more deeply rooted are self-managing teams; they represent the transfer of considerable authority down the corporate ladder. Teams have authority over both "doing things right" and "doing right things." They also determine the composition of the team itself; they can hire and fire. Not surprising, this requires extensive training. Once self-managing teams are established, there is also no turning back.

Managers get a humane role... With any approach that takes advantage of 100 percent involvement, a new role is offered to managers; that of resource, teacher, leader, and coach. It's a better role than that of minor autocrat or, worse yet, being a conduit for decisions reached at a higher level; but it is a new role and requires preparation.

QUALITY TAKES ROOT IN STAGES

Healthy growth is possible only when the process is well-defined and employees are prepared. A "Quality by Proclamation" approach—when a CEO announces that his or her organization is a "Quality Company" and the major expense and effort is devoted to advertising the new corporate image–is an example of trying to establish a lot of ground cover with no roots. It won't survive the first crisis.

Setting up a quality process and standards True commitment—active, obvious, and informed commitment – is attained only when top management is willing to go through the tough work of defining exactly what the mechanics will be for achieving quality in their organization. Then employees have to be educated about the process and how they can contribute. Naturally, if the process involves 100 percent of the employees, growth is far faster and covers a lot more ground.

As quality spreads, it begins to take over everything the organization does. It becomes a part of the organizational vocabulary; it literally pervades every decision taken at every level. Companies eventually encourage their neighbors to begin to focus on quality through mutually beneficial quality partnerships. (No one wants to import crab grass from a neighbor.)

Over time, customers can and do distinguish between companies that claim quality and those that produce quality. It is equally true that the benefits of quality are available to any organization willing to do the work. There is no reason why the grass must always be greener on the other side of the corporate fence.

Journal for Quality and Participation, *Sept. 1990*

THE 100% SOLUTION OR GREATER NON-CONFORMITY

The approach of the 21st century is bringing one thing sharply into focus: the century will be about maximums, not minimums. Organizations that meet ISO 9000 standards will be in a position to be players; organizations that meet the Baldrige criteria will be in a position to be winners. Organizations that do things will survive; organizations that do the right things – often as yet unimagined – will thrive.

Nowhere is the challenge facing leading-edge organizations greater than the need to evolve to a vision of the future which includes the infusion of 100 percent of each employee into the workplace.

Historically, this represents a reversal of centuries of tradition, going beyond timid employee involvement programs currently fashionable. (Which is not to say that most current employee involvement programs are not an improvement on the past.) Beginning with the old guild system, youths were apprenticed for their labor, nothing else. They were there to learn. Throughout the long course of apprenticeship and journeyman, new ways of doing things were undelivered. Even a *masterpiece* was merely a way to demonstrate a grasp of the minimums.

The industrial revolution exacerbated the situation. Machines were necessary to meet the growing demands of an awakening economy. Basic skills, preferably simple and easily acquired on the job, were needed; people who were interchangeable were required. How to get a large number of interchangeable people?

If an organization had *A*, *B*, *C* and *D* as employees, by knocking off, or ignoring all unwanted attributes, it gets a set of new people with different features.

It was neat, it was predictable, and if Person *B* didn't happen to work out, the boss could easily reach back into the pool of employees and substitute another of the uniformly talented workers. Job recruitment was simplified; the perfect recruit was someone who most closely fit the demands.

Fitting into the box is the Ideal... Even in situations where more complex skills were required, like mathematics or reading or writing, extraordinary skills of any sort were considered undesirable. Most pre-screening systems were designed to ensure that the prospect's talents were sufficient for the task at hand—and only that. The number of people who have been turned down for jobs because they were *overqualified* was—and regrettably still is—legion.

This approach to managing workers (*that we can replace any of them/no thinking allowed theory of management*) did produce merchandise, but it also gave rise to adversarial labor-management relations. And it encouraged individuals to disengage themselves emotionally from their workplace.

As a result, even 100 percent employee involvement, one of the foundations of the quality revolution, was considered a pipe dream less than 20 years ago. Everyone knew that management knew best, and workers existed to carry out management directives. Frederick Taylor made that abundantly clear.

Participation, why, yes—sure! The notion of the importance of employee participation slowly gained acceptance for the most straightforward of reasons: it made solid economic sense. If there are 2,000 people on the payroll, why devise a structure that only asks 200 of them to contribute to the improvement of the organization while the talents of 1,800 go unused—or at least under-utilized? The quality revolution of the 1980s predicated putting management and non-management on a more equal footing regarding day-to-day decision-making within an organization. Too often, however, that drive took a peculiar twist.

It is an errorless process or is it airless? In many organizations, with the need for 100 percent participation accepted, processes were devised under the best circumstances (using the talents of a broad spectrum of employees), and then the processes themselves gained ascendancy over people. With the correct processes in place, it was felt nothing could go wrong—and everyone ceased to think, management and non-management alike. Equal footing, indeed, albeit a reversal of the original intent. The lack of dynamism in such an approach to quality improvement is now being reflected in complaints that quality does not encompass creativity and growth strategies. In point of fact, many programs labelled TQM don't. When a program's main focus is to ensure that nothing can go wrong, it does not constitute the best that the quality revolution has to offer.

Putting people, individuals first... Thankfully, some organizations have taken more ambitious approaches to quality. These organizations look for ways to put people first, look for ways to engage and engross employees in their work—and ask them to dream of possibilities as yet unenvisioned. In short, some organizations already ask for 100 percent of each individual.

This is a more subtle concept with the same commonsense foundation found above: if you are going to pay someone to spend eight or more hours taking part in the activity of your organization, why not attempt to take full advantage of all the talents of that individual?

The benefits to an organization are obvious. Consider the case where a uniquely talented individual is aligned with others who are equally idiosyncratic.

The final outcome is roughly a combination of the above two. The example is, of course, simplistic. The normal case would have gaps between skill and/or overlap. The choice, though, is to either go back to the second case where skills were altered, which intentionally limits an individual's contribution, or to find a way to fill the gaps and use the rest.

Can the real world support the 100% solution? Ironically, filling the gaps may be more difficult in the immediate future. Just as this second major revolution is about to begin, i.e., the move toward infusing 100 percent of each employee into the workplace, our educational system's contribution is being called into question.

Are new workers coming to the job market with the basic skills intact? In one court decision (over seven years ago) a major employer was ordered to give remedial classes in English and mathematics to make up what had not been taught—or at least learned—in high school by its potential employees. Virtually every day, new horror stories about the educational system find their way into the newspapers, one of the latest being about the mother of a third-grader complaining because her child's homework was too difficult.

The need for continual education... This is where the concept of continuous education that Tom Peters, among others, has been touting comes into play. Pioneers such as Johnsonville Sausage, for instance, gave automatic pay raises for the completion of any education course by any of their employees—whether or not the course has a direct link to the worker's job.

The goal is continuous education. The theory is that if people's brains are kept active, all aspects of their lives benefit.

And the rewards to the individuals, their leaders, and the organization as a whole, make such investments more than worthwhile.

In the 21st century, a new factor will be added to the cost of non-conformance. Savvy organizations will come to appreciate that the expense of wasted talent may well exceed the combined cost of prevention, detection, correction, and failure.

Journal for Quality and Participation, *Dec. 1993.*

5

MEASUREMENT

"Measurement: Neither a Religion nor a Weapon" describes the vital role of measurement in a quality process, with the story of Manager B and Employee A serving as a reminder that measurement is not entirely a rational exercise. Even when hard, provable data is available, human emotions and relationships determine how useful the numbers really are.

The other two articles in this chapter give examples of how measurement might be used to assess the worth of the quality effort itself. "Sharing the Wealth in Quality Partnerships" demonstrates how a partnership between two organizations can be to their mutual benefit and to their customer's benefit as well. It is based on the formal partnering between organizations introduced into the American business vocabulary and practices by Motorola after it won the United States Malcolm Baldrige National Quality Award in 1988.

"Qualicrats and Hypocrites: A Troubling Status Report from the Front" describes a short survey given to the management team of an organization that prided itself on being a "quality organization." When the results were negative, the team was, not surprisingly, displeased.

MEASUREMENT: NEITHER A RELIGION NOR A WEAPON

Any discussion of measurement in the context of a quality effort needs to begin with the refutation of a myth and the establishment of a guideline. The myth, a peculiarly tenacious one, is that "The one who dies with the most charts wins." This simply is not true. It is surprising, despite dozens of examples of quality efforts that have drowned in paperwork, that so many executives still confuse *volume* with *value* when it comes to statistics. These same executives are also likely to be dazzled by complexity in number taking. Remember this valuable (although not voluminous) guideline: "If you have to take a square root, you've gone too far."

In brief, too often measurement is overdone and overcomplicated. Call it the "religion" approach to measurement, absolute belief that numbers are sacrosanct and that the ritual of taking numbers is a higher calling. Few people actually believe this; many people behave as if they do.

The American attitude toward statistics is paradoxical: we do not like statistics, but we take a great many of them. On one hand, one of the first truisms memorized by school children is Mark Twain's "Lies, damn lies, and statistics"; at the same time, statistics ranging from batting averages to academic standings to weekly and gross sales for movies creep into our everyday conversations.

This paradox extends into business. If a survey were taken to determine which of the three major components of a quality process—leadership, participation, or measurement—was the least popular in the sense of being liked or enjoyed, measurement would surely be the runaway winner. Yet if a study were made of quality efforts launched in America during the last decade to determine which of the three is most popular in the sense of being most often emphasized, measurement would again be the winner.

The gathering of numbers is, in and of itself, a neutral occupation. It is neither good nor bad. Only when numbers are put into use and a social context given to them do emotions (besides perhaps tedium) come into play. Unfortunately, measurement generates a lot of negative emotion, even when the goal is quality. This is not to be wondered at when the common approach to defining which measurements get taken often begins with the statement. "If things don't get better around here soon, we're going to start measuring."

True, numbers are used to identify and solve problems. Also true, there are a number of ways to present those solutions. Imagine for a moment that Employee A cooperates with Manager B in the recording of a specific piece of data over a period of a month. At the end of the month, Manager B confronts Employee A with the statement, "Aha! I've finally figured out what you are doing wrong that is costing us so much time and money. Now I can make sure you don't hurt us any more." Any resolution coming out of that discussion will likely be adversarial and temporary.

If Manager B, however, approached Employee A and said, "These numbers have me bothered. Take a look at them and I think you'll see that we still aren't getting the results that we thought we would. I'd really like to get your reaction. Any ideas on how we can make things better?" any resolution coming out of their discussion would have a much better chance of creating teamwork and lasting improvement.

In the first case, Manager B was using the data as a weapon, an all-too-common example of the misuse of measurement. When measurement is used in this way, short-term fixes (if any) and long-term resentment (assured) are the result. It would be just as useful to roll up the report and jab the employee in the eye with it. The data-as-a-weapon approach to measurement leads to some odd scenarios. Consider the following story, sent to us in August 1992 from a mid-level executive with one of America's largest multinational firms:

"In our training we were told about control charts that graph samples and their deviation from the desired values, showing a process is 'out of control.' We were shown several patterns of data and asked what conclusions we would draw. For example, one set showed samples within the upper and lower control limits but steadily declining. My favorite chart showed samples alternating between above and below the desired values. My reaction was that the chart showed some regular phenomenon (as if the process was to cut a metal bar into two equal pieces, with one always ending up too long and the other too short). I was in the minority. The first person to speak said that the chart looked too perfect; he'd assume that someone was rigging the data. The instructor agreed; his reaction would be that the employees were reporting false data.

As we were about to go on, I volunteered that my first reaction would not be to distrust my employees, but to look at the data. That view didn't get many supporters, though."

It is apparent that the managers who assumed that the employees were lying and that the data was right subscribe to a corporate culture in which the presumption of innocence (or good will) on the part of a subordinate is not an option. Even if a state of open warfare between the "classes" does not yet exist, it is a good bet that guerilla warfare is constant, and that process improvement, despite reams of data, is erratic at best. It is true that punishment and false data are linked but the managers in the example have the sequence backward.

The moment when data begins to be used to punish people is normally about 60 seconds before those being measured start looking for ways to beat the system—including falsifying data. Although the timing is tight, the sequence is important and underlines the error in thinking in the example. Employees falsify data to stay out of trouble, not to get into trouble. When numbers are primarily used for finding the "guilty party," the one thing that every employee who will be measured knows as an article of faith is that an out-of-control process means trouble.

For companies hoping to improve their processes, services, and products, there are only two appropriate uses of measurements: as a source of ideas and/or a means for tracking progress. Used in one of these two ways, measurements are a rich vein of information for devising improvements. Even if a process is already meeting or exceeding specifications, the information can be used to move to an even better level of performance. See if you can spot what is wrong with this approach: Manager A approaches Manager B with the news that they have decided to measure 50 parameters on their common business process and then decide how they can best use the numbers. How does that sound to you?

We are back to measurement as a religion. Not every number is useful. Why take numbers in the first place if you don't have something specific in mind? Choosing what to measure is at least as important as the actual measurements themselves. The correct criteria is not, "What would be easy to measure?" but rather, "What would be of use to know?" The former question is the first step toward collecting charts by the hundreds; the latter can lead to the determination of a measurement plan, a far more appropriate and economical activity.

Any good plan includes "tripwires." Anyone who has been to a Boy (or Girl) Scout Jamboree (or other large gathering of young people in some out-of-doors setting) is familiar with the concept of tripwires. In scouting, it is a time honored tradition for one troop to "raid" another in the middle of night (in search of a flag or some other trophy or just for simple harassment purposes). It is usually not possible to patrol the entire perimeter of each position throughout the night, so tripwires are set. These are wires or strings or ropes with noisemakers attached, placed so that any raiding party will trip across them.

The idea is that the resultant clatter will awaken the "home team" in time to turn back the intruders. The trip-wire itself does not stop anyone; it does indicate where problems are arising and resources should be expended.

In the world of business, a tripwire is any measurement which indicates that more specific measurements need to be taken, i.e., where an organization should concentrate its limited resources. An example would be the current airline statistics on lost baggage. So long as an airline's lost baggage rates are in the top two or three for the industry, normal continuous improvement and managerial attention can be assumed to be doing the job. If, however, the numbers begin to indicate problems, then that airline's managers know that special attention needs to be paid to the process. Additional measurements can be put in place and the data collected used as a source of ideas. Once the process reaches the point where it is securely on track once again, the number of measurements can be reduced and the resources can be redeployed. But what to measure? Particularly in service industries, many individuals attempt to excuse themselves from measurement by stating, "You can't measure what I do." If that were actually true, on what is their compensation to be based?

A beginning point for determining measurements that will be of value is the question, "How do you know when you've done a good job?" If that question draws an initial blank, how about, "What makes your customer think you've done a good job?" or "How does your boss know when you've done a good job?"

Such questions will, of course, be met with suspicious truculence if the track record of the company is to use measurements as a means of punishment. Only if everyone is convinced that measurements are to be used only as a source of ideas—to fix processes not people—and as a way to track progress will cooperation be forthcoming.

Assuming that the groundwork has been laid and employees are convinced of the efficacy and desirability of actually gathering and interpreting statistics, there is one more hurdle to overcome. Traditionally, measurement tools have acquired an aura of mystery. Consider SPC for a minute. Its reputation for sophisticated manipulation of data is left over from the time (15 or more years ago) when the value of the whole idea of quality was grossly underestimated. In a classic job-saving move, quality control professionals encouraged others in the belief that SPC was difficult, surmising that while management might fire someone whose profession was not all that complicated (since it would be easy to re-build the capability if ever needed), management was not likely to fire someone whose job they didn't understand.

It is time for a little heresy. The science of measurement—often summarized under the collective title "Statistical Process Control"—simply isn't all that complicated. For the most part, it consists of addition and subtraction. For the multiplication and division that is needed, a calculator makes it easy. And, if you are called upon to take a square root, refer to the first paragraph of this article.

It is the '90s now and it is time for SPC and its practitioners to drop the mystique. Take the Pareto Analysis, for example. This has become the crown jewel of SPC and has grown to such importance that it has become a verb, as in "Let's Pareto that problem."

The entire theoretical basis for the Pareto Analysis is that an Italian named Pareto noticed one day that a few folks have most of the money. This became the 80–20 rule, was one day chiseled in concrete somewhere, and is now a Given Truth. It is a powerful and versatile tool that every manager should be familiar with—but all it really amounts to is, "Let's keep track of what's going on for a while and then work first on the thing that is causing the most trouble." This is not nuclear science.

Measurement, to include SPC, is an absolute necessity for the implementation of changes in or improvements to any business process. It is, however, only one of the three major components of a balanced, effective quality process. The other areas are also covered in *Quality in Action: 93 Lessons in Leadership, Participation, and Measurement.*

Measurement is last for a very good reason. It is only effective when a management team has already articulated a vision, defined a structure that enables every person in the organization to participate, and modeled quality improvement in the performance of their own jobs. Measurement, however, is not least. To emphasize any of the three components at the expense of either or both of the others is to doom a quality process. Success—sustained, continuous success that not only boosts the bottom line, but also ensures the loyalty of customers and raises the morale of employees—is possible only if leadership, participation, and measurement are all present.

The Quality Observer, *Oct. 1992.*

SHARING THE WEALTH IN QUALITY PARTNERSHIPS

If there has been one overriding business lesson in the last two decades, it is that quality pays. While, historically, organizations began quality improvements within their own corporate structures, quality partnerships between customers and suppliers developed as a way to tap additional savings for both parties. When those relationships were true partnerships–rather than thinly veiled attempts on the part of large customers to extort savings from smaller suppliers–these partnerships worked. The following hypothetical example illustrates how.

Company S (supplier) is a relatively small company with the reputation of being a reliable widget maker. It has several worldwide competitors. Company C (customer) uses Company S's widgets to manufacture a larger, more complex product.

It costs Company S $1 to make each widget. It sells the widget to Company C for $ 1.07. Company C actually buys widgets from several different suppliers because, in the past, shipments from suppliers have been slightly erratic and a certain percentage of widgets have been defective. Figuring in inventory costs and the cost to replace defective widgets, Company C's real cost for each widget is $ 1.12.

The marketplace starts to get tough for Company C, so its leaders tell employees to do a better job and to save money. In response, someone suggests that the widget suppliers be sent a letter stating, "We at Company C believe in quality. As your biggest customer, we think you should, too. Incidentally, we'll be expecting your price to drop by 5% in the next year."

Cooler, wiser heads prevail, and they scrap the idea. They realize that this one-way ultimatum will potentially cut the supplier's profit margin and cause the working relationship between Company C and its suppliers to deteriorate. Instead, Company C sends letters to its suppliers inviting them to become quality partners.

Only one supplier, Company S, decides to take Company C up on its offer. At a meeting, they agree to develop a quality partnership. Since Company C has had some experience with quality (several people are familiar with quality tools and concepts) and has deeper pockets, it offers to fund the needed quality training.

Key employees from Company S attend Company C's training, and the companies share data that were once considered sacred. As in many quality partnerships, many visits are made back and forth. At first, executives and managers go on these visits, but then those who actually work with the widgets go. Not surprisingly, these workers find opportunities to improve the widget and reduce costs. Statements such as "Did you realize that if you store the widget like that, you're going to damage the bottom layer?" are common.

After a year, both companies are benefiting from the partnership. Company S has reduced its cost to produce widgets from $1 to 90 cents. Company C is no longer worried about delivery schedules and defective widgets since the two companies are working closely together. In addition, Company C now can buy widgets solely from Company S when it needs them, reducing inventory costs from 5 cents per widget to 1 cent per widget.

Company C and Company S agree to a new price of $1.02 per widget (formerly $1.07). Company S is now making a profit of 12 cents per widget vs. the previous profit of 7 cents—and making that profit more often since its volume of sales to Company C has increased. As word of the widget's improved quality and reliability spreads in the marketplace, sales to other customers also increase.

Company C is now paying $1.02 for each widget, plus an additional penny per widget for internal costs, so its real price per widget has been reduced by 9 cents (from $1.12 to $1.03).

As the result of the quality partnerships, Company S is making more money and Company C is spending less. But is there any downside? It depends on how one looks at it. For Company C, it must be admitted that not all of the savings accrue to the bottom line. Initially, Company C has to absorb the cost of training and other costs incurred in initiating the quality process.

Another downside could occur at Company S. As Company S expands its capacity through increased efficiency, it has two choices. If it is shortsighted, some employees might lose their jobs. If it takes a longer view, however, it will take advantage of the early gains in productivity by enabling employees with "free time" to partake in cross-training or in new-skills training. By doing so, Company S will prepare for the increase in business that will inevitably result from the improved quality and reliability.

What would have happened if Company S had improved quality without help from Company C? On the downside, it would have had to perform the basic research alone, fund its own training, and pay all travel expenses for visits to Company C. On the upside, it could have per haps pocketed all of the increased profit margin instead of sharing it. At the least, Company S would have had a lot more room to bargain in competitive situations, especially if its competitors were working within the previous budget and profit margin constraints.

As this example shows, both the supplier and customer benefit from a quality partnership. The only real losers in this scenario are Company S's competitors. They ignored Company C's invitation to become quality partners, and now they will find themselves with reduced sales and relatively small profit margins.

Quality Progress, *July 1994.*

REFERENCE

In the average U.S. manufacturing company, 20% or more of sales income is used to cover non-conformance costs (i.e., the cost to prevent, detect, and correct errors combined with the cost of lost business due to failures that get out the door). For the average US service company, the figure rises to 35% of operating costs (Philip B. Crosby, *Quality Is Free*, McGraw-Hill Books Co., New York, NY, 1979). Even if Company S had made previous quality attempts to economize in the past, experience shows that some savings can be realized—unless, of course, its processes and products were already perfect.

QUALICRATS[1] AND HYPOCRITES: A TROUBLING STATUS REPORT FROM THE FRONT

While neither qualicracy[2] nor *qualicrat* is in any current dictionary, this article is written to express the fear that their inclusion is only a matter of time. It draws from an experience with a prime example of qualicracy: an organization whose quality programs were built by following the guidance of high-profile consultant companies (whose prices were inversely proportional to their trust in the capabilities of the average employee), and which came complete with senior executives who have adopted the mantra, “We've already done quality. It is so much a part of who we are that we have no need to formally assess quality on a day-to-day basis.” This unnamed organization has won recognition on several occasions at the state and federal level for its quality and productivity efforts. It is a unit of the federal government that provides services to a wide range of other federal units. But, just as all bureaucracies are not governmental, neither are all qualicracies. While Uncle Sam has become the largest single purchaser of qualicratic efforts, he is by no means alone in the practice of being better at self-congratulations than at achieving measurable results when it comes to continual improvement.

A survey was administered to provide information for an after-dinner speech at the conclusion of the first day of this organization's annual conference on organizational

[1]*Qualicrat:* 1. An official who behaves as a bureaucrat while professing the value of blending customer satisfaction, leadership, authority equal to responsibility, trust, communication, training, recognition, and measurement. 2. A hypocrite.

[2]*Qualicracy:* Quality vocabulary grafted onto bureaucracy, marked by inattention to individuals; an administrative system in which the need to follow complex procedures impedes effective action, marked by a lack of measurable results.

excellence, a three-day affair whose attendees were the 33 senior managers of the organization. Under the assumption that shared knowledge is the hallmark of a quality organization, the survey was intended to gather specifics that the speaker could use to congratulate the managers on their successes and to offer suggestions for incremental improvement of current procedures. The survey took about five minutes to complete, but managers were soon aware that the results conveyed substantial disagreement about what the organization had achieved and at what cost—a disconnect severe enough to disturb the celebratory atmosphere.

Figure 1 gives the 33 managers' responses to the nine survey questions. Note that the responses are numbered for ordering purposes only; each number does not correspond to a particular individual (i.e., the person who gave answer 1 in question 1 is not necessarily the same person who gave answer 1 in question 2).

1. *Expressed in dollars, what has been the average annual impact on the bottom line of [this organization's] quality efforts over the last three years?*

The elicited answers —which ranged from zero to $1.2 billion—indicated that calculating the dollar impact of an idea prior to or after implementation was not a common practice. (During the speech, when these results were presented, the head of the organization ventured that it was difficult for a federal body to measure dollar impact.) Not only were monetary savings not used as a measure of improvement, alternative methods of evaluating progress were also absent. There weren't even data on time savings. The speaker's subsequent conversations with a senior executive revealed that customer survey data were as scarce as savings data. The senior managers (particularly the top few) were positive that they oversaw a first-rate quality operation, but they failed W. Edwards Deming's most basic requirement: "Show me the data."

Whether a certain organization has built a quality process according to a particular set of guidelines is not the definitive assessment of its worth. Results count. But when no data are available, it is necessary to look elsewhere. The level of trust in, and knowledge of, employees on the part of senior executives can offer a revealing look into the soul of an organization and, by implication, its long-term effectiveness and prospects.

2. *As of today, how many of* [*the organization's*] *approximately 2,000 people are actively, formally engaged in some aspect of the organization's quality efforts*?

While most quality organizations have a rough idea of who is working on quality-related improvement, a qualicracy usually does not. In answering question 2, there was a striking lack of agreement among the respondents. When the speaker asked about the wide range of answers to this question, the response was, "We've told all of our people that they are responsible for quality all the time." Further, survey respondents objected to the phrase "actively, formally engaged in some aspect of the organization's quality efforts," saying that it was too vague.

During a visit to the organization a week before the conference, the speaker was told of a recent recognition luncheon that had been given for employees on active Project Action Teams. A typical feature of a qualicratic environment, such teams are normally made up of 10 or so non-managers and are overseen, guided, chartered, and managed by two to four tiers of boards, committees, and groups made up of managers. Since 126 team members were honored, the speaker's expectation was that the answers to question 2 would range from 100 to 150 (roughly one in five of the responses fell in that range), giving him an opportunity to suggest in his speech that the organization move toward 100% formal involvement.

3. *As of today, how many of* [*the organization's*] *approximately 2,000 people would you estimate have personally contributed an idea for quality improvement in the last five years*?

With the responses to this question, the low level of involvement became even clearer. An average of 6.7% of the workforce contributed ideas for improvement in any given year, an unimpressive number.

4. *How many hours of formal quality-specific training have you personally received?*

Although there is little doubt that a great deal of time and money has been spent on training, other survey responses raise questions about its usefulness. And these answers raise serious questions about the structure of a training program (or, perhaps, senior managers' commitment to it) in which the levels of training on a single topic could vary so dramatically.

5. *Approximately what is [the organization's] annual training budget?*

In most organizations, the size of the training budget is virtually unknown, so this question was asked with no expectation of consistent answers, but rather to set up a discussion around the next question concerning quality-specific training. The surprise in question 5 was the extent of common knowledge. Twenty-eight of the 33 responses fell in the $1.9-million-to-$3-million range, with 10 responses clustered between $2.6 million and $2.7 million. As it turned out, it is a point of pride in this organization that more than $2.5 million is devoted to training each year from an annual operating budget of $1.2 billion, so the number is advertised widely. In short, when a number is considered worthy of note, the senior management team does communicate reasonably well (which makes the answers to question 1 all the more disappointing). In point of fact, the total outlay for training is 2% of the budget, one half of the 4% currently thought necessary to remain competitive in the private sector.

6. *Approximately what percentage of [the organization's] annual training budget is used for quality-specific training?*

This question produced little consensus, but no uniformity was expected. This is the kind of information rarely known within organizations, even at the senior management level. The intent of the question was to provide a springboard for discussion on an appropriate expenditure for quality, with the reminder that Motorola determined the return on investment for its quality-specific training to be 30-to-1. The term "quality-specific" appeared to cause confusion among the respondents: one manager stated, "Well, I could argue that any training I go through is quality-specific." With the responses to question 4—time spent in quality-specific training—ranging from eight hours to more than 1,200 hours and estimates of the cost varying wildly, it is impossible to get a clear picture of who is being trained to do what and at what cost.

7. *If your answer were kept in confidence, how many employees at* [*this organization*] *could you name who you think are incapable of contributing to the improvement of their own job or the organization's operations*?

Admittedly somewhat awkward, question 7 was posed to gain some idea of how much trust and confidence the organization placed in its employees. Another way this question could have been worded would have been, "If you were to begin your quality effort today, how many people do you think you could afford to exclude?" Unless the respondents were all referring to the same person or people, there are at least several dozen people on the payroll who are thought to be unable to contribute to the improvement of their jobs or the organization's operations. Whether the respondents' opinions are based on experience or are indicative of a lack of respect or a lack of trust is a matter of guesswork.

The senior managers' response to the presentation—assuring the speaker that quality efforts were both active and sophisticated—contrasted sharply with the organization's moribund structure for utilizing employees' talents. The easiest way to spot a qualicracy is by applying the criterion developed by Texas quality observer Keith Taylor: "Are their hips and their lips going the same direction?" In this case, the answer is no. While most senior managers implied that they have a high regard for all but a relatively small number of employees, their responses to a later question suggest that this alleged trust does not translate into quality improvement activity. Two offhand comments suggested why this might be so. The week before the conference, while discussing quality efforts, one senior manager—a true qualicrat—informed the speaker, "We have a lot of people who aren't capable of doing this." After taking the survey, however, this same invididual laughingly told another manager and the speaker that he had answered "zero" to question 7 because he knew that was the right answer. Employees have particularly keen noses for such hypocrisy. Trust vanishes at the first whiff.

8. *As a percentage of the annual budget, what do you estimate* [*the organization's*] *cost of quality to be*?

A definition of the cost of quality was included in question 8, but because this organization's customers effectively have no other suppliers, the definition did not include the element that Deming held to be the largest single component of quality cost: lost customers. (Traditionally, cost-of-quality calculations ignore another component that directly affects the bottom line—not tapping into the abilities of every person on the payroll.) When asked to pinpoint the cost of quality, defined as prevention, detection, correction, and the cost of uncorrected errors in the organization, the answers ranged from 0.5% to 100%. Such a wide range might be acceptable from a random group, but coming from the senior management team of an award-winning organization, these answers simply add to the speculation about what exactly was taught in all those hundreds of hours of quality training. Only five respondents' answers were in the range generally acknowledged to be applicable to service enterprises: 30% to 35%. It is possible to lower the cost of quality by shifting attention and costs to prevention, but no one raised that point during the discussion, suggesting that no one knew whether that was the organization's strategy.

9. *If an employee were to come to work with an idea that he or she believes would save the equivalent of one person's time for 20 minutes a week (every week from now on), what should he or she do with the idea? Assuming it is a valid idea, approximately how long would it take to have the idea formally implemented?*

The most disconcerting answers came in response to this question. Six of the responses to the first part of question 9 were blank. Other answers to the first part of the question clustered around some use of the suggestion system.

1. Expressed in dollars, what has been the average annual impact on the bottom line of [this organization's] quality efforts over the last three years?

1. 0	12. $150,000	23. $300 million
2. 0	13. $300,000	24. $1.2 billion
3. $1,000	14. $1 million	25. 2%
4. $1,000	15. $1.2 million	26. 15%?
5. $15,000	16. $10 million	27. ?
6. $50,000	17. $10 million	28. –
7. $100,000	18. $10 million	29. –
8. $100,000	19. $18 million	30. New to organization-no feel
9. $100,000	20. $25 million	31. Unknown–$10 million
10. $100,000	21. $30 million	32. Unknown—less than $10 million
11. $150,000	22. $100 million	33. Don't know

2. As of today, how many of [the organization's] approximately 2,000 people are actively, formally engaged in some aspect of the organization's quality efforts?

1. 1	12. 400	23. 1,000
2. 20	13. 400	24. 1,000
3. 25	14. 400	25. 1,400
4. 40	15. 400	26. 1,500
5. 100	16. 500	27. 1,500
6. 100	17. 500	28. 1,600

7. 200	18. 1,000	29. 1,600
8. 200	19. 1,000	30. 1,900
9. 200	20. 1,000	31. 2,000
10. 200	21. 1,000	32. 2,000
11. 250	22. 1,000	33. Define quality efforts

3. As of today, how many of [the organization's] approximately 2,000 people would you estimate have personally contributed an idea for improvement in the quality of some aspect of [the organization's] operations at some point in the last five years?

1. 50	12. 200	23. 1,000
2. 100	13. 250	24. 1,000
3. 100	14. 250	25. 1,200
4. 150	15. 300	26. 1,500
5. 150	16. 400	27. 1,900
6. 200	17. 400	28. 1,900
7. 200	18. 400	29. 1,990
8. 200	19. 423	30. 2,000
9. 200	20. 500	31. 2,000
10. 200	21. 600	32. –
11. 200	22. 1,000	33. –

4. How many hours of formal quality-specific training have you personally ever received?

1. 8	12. 80	23. 120*
2. Less than 10	13. 80	24. 160
3. 10	14. 80	25. 200
4. Less than 20	15. 80	26. 200
5. 20	16. 88	27. 200
6. 40	17. 100	28. 200
7. 40	18. 100	29. 200*
8. 40	19. 100	30. 500
9. 60	20. 100	31. 1,000
10. 60	21. 100*	32. 1,000*
11. 60	22. 120 (too much!)	33. 1,200

* Where possible, percentage answers were converted to numerical answers.

5. Approximately what is [the organization's] training budget?

1. $250,000	12. $2 million	23. $2.6 million
2. $500,000	13. $2 million	24. $2.7 million
3. $1 million	14. $2 million	25. $2.7 million
4. $1.6 million	15. $2.1 million	26. $2.7 million
5. $1.9 million	16. $2.3 million	27. $2.8 million
6. $1.9 million	17. $2.6 million	28. $2.9 million
7. $2 million	18. $2.6 million	29. $2.9 million
8. $2 million	19. $2.6 million	30. $3 million
9. $2 million	20. $2.6 million	31. $3 million
10. $2 million	21. $2.6 million	32. $3 million
11. $2 million	22. $2.6 million	33. $10 million

6. Approximately what percentage of [the organization's] annual training budget is used for quality-specific training?

1. 0.1%	12. 10%	23. 50%
2. 0.2%	13. 10%	24. 100%
3. 0.5%	14. 10%	25. $50,000
4. 1.3%	15. 10%	26. $90,000
5. 5%	16. 10%	27. $100,000
6. 5%	17. 15%	28. $100,000
7. 5%	18. 25%	29. $100,000
8. 5%	19. 25%	30. $300,000
9. 5%	20. 25%	31. $300,000
10. 5%	21. 30%	32. $900,000
11. 10%	22. 50%	33. $1 million (too much)

7. If you were guaranteed that your answer would be kept in confidence, how many specific employees at [this organization] could you name who you think are intellectually incapable and/or emotionally unable to contribute to the improvement of some aspect of their own job or of [the organization's] operations?

1. 0	12. 0	23. 10
2. 0	13. 0 or few	24. 10
3. 0	14. 1	25. 10
4. 0	15. 1	26. 20

5. 0
6. 0
7. 0
8. 0
9. 0
10. 0
11. 0
16. 1
17. 3
18. 5
19. 5
20. 5
21. 5
22. 3 to 10
27. 20
28. 25
29. 25
30. 90
31. 100
32. –
33. Too many/can only name a few/too few/ maybe 10% to 12% (200 to 240)

8. Quality theorists talk about a value called "cost of quality." It is usually expressed as a percentage of an organization's annual budget and is said to represent the total of the cost to prevent problems plus the cost to detect problems plus the cost to correct problems plus the cost of uncorrected error. As a percentage of the annual budget, what do you estimate [the organization's] cost of quality to be?

1. 0.5%
2. 1%
3. 1%
4. 3%
5. 5%
6. 5%
7. 5%
8. 5%
9. 5%
10. 10%
11. 10%
12. 10%
13. 10%
14. 15%
15. 15%
16. 15%
17. 20%
18. 25%
19. 30%
20. 30%
21. 35%
22. 35%
23. 35%
24. 40%
25. 45%
26. 50%
27. 50%
28. 60%
29. 70%
30. 100%
31. $30 million
32. $100 million
33. –

9. If an [organization] employee—at any level—were to come to work next Monday morning with an idea that he or she believes would, if implemented, save the equivalent of one person's time for 20 minutes a week (every week from now on), what should he or she do with the idea? Assuming it is a valid idea, approximately how long would it take to have the idea formally implemented?

What should the person do with the idea?

1. Use the organization suggestion system.
2. Use the organization suggestion system.
3. Use the organization suggestion system.
4. Use the organization suggestion system.
5. Use the organization suggestion system.
6. Use the organization suggestion system.
7. Use the organization suggestion system.
8. Use the organization suggestion system.
9. Use the organization suggestion system.
10. Use the organization suggestion system.
11. Use the organization suggestion system.
12. Use the organization suggestion system.
13. Submit/tell someone.
14. Elevate to immediate supervisor.
15. Vocalize.
16. Present it through various channels such as management suggestion program.
17. Verbalize it to proper person.
18. Act or see leader or use the suggestion system.
19. Suggestion system plus talk to supervisor to implement locally.
20. Use the suggestion system or discuss with supervisor or person responsible for implementation.
21. Implement it themselves or use suggestion system.
22. Bubble up to management.
23. Implement.
24. Do it.
25. Push it.
26. Implement and tell boss later.
27. Forget it.
28. –
29. –
30. –
31. –
32. –
33. –

How long would it take to implement the idea?	
1. Same day	17. 6 months
2. Immediately to 6 months	18. 6 months
	19. 6 months
3. Immediately to never (depending on branch)	20. 6 months
	21. 6 months
	22. 6 months (maximum)
4. 2 weeks	23. 6 months to 2 years (if it is a systems change)
5. 30 days	
6. 30 days	
7. 1 month	24. 1 year
8. 45 days	25. 1 year
9. 1 to 2 months	26. 1 year
10. 2 months	27. 1 year
11. 2 months	28. 2 years
12. 2 to 3 months	29. 2 years
13. 3 months	30. 2 years
14. 2 to 6 months	31. 5 years
15. 6 months	32. Probably never
16. 6 months	33. Depends on complexity

Figure 5.1 Survey questions and responses.

One of the trademarks of an organization with a functioning quality process is that the question, "What do you do with an idea, large or small?" will elicit a consistent response, no matter who one asks. Although the suggestion system in this organization is highly touted, there were fewer than 200 suggestions submitted in 1994 by the 2,000 employees. By contrast, Milliken, a Malcolm Baldrige National Quality Award winner in 1989, got 67 ideas *per employee* in 1994. Apparently, no accurate records are kept at this organization of the ideas submitted to supervisors or ideas implemented. The need for record keeping, of course, is so that each succeeding idea can build on the platform created by previous success. It is the accumulation of knowledge that makes continual improvement possible.

Looking at how long respondents said it takes to implement an idea (see Figure 5.1) gives some indication of why employees don't use the current system with any frequency. Most of the senior managers believed it is acceptable to take six months or more to implement even a small idea. Their responses are a testament to the qualicratic nature of the organization's efforts at continual improvement and evidence of why so many employees—the frontline combatants in the American quality revolution—are tiring of the fight.

It should come as no surprise that the after-dinner speech was not a success, although several of the junior members of the audience sought out the speaker and thanked him—but only after their senior managers had left. Their furtive enthusiasm intensified the speaker's sense of foreboding: A shared vision of the future is made particularly difficult if there is no shared understanding of the past and present.

While the nine questions in this survey hardly constitute a definitive test of the quality efforts of any organization, they can serve as a quick check of the common knowledge held by any subset of an employee population. Unclear or alarming answers can alert management to pay particular attention to, and conduct further investigation into, specific areas. So long as organizations ignore or misuse the talents of their employees–as the organization profiled here clearly does–American quality will continue to fall short of its potential.

Quality Progress, *Jan. 1997.*

REFERENCES

1. Marion Steeples, *The Corporate Guide to the Malcolm Baldrige National Quality Award* (Milwaukee: WI: ASQC Quality Press, 1993), p. 248.
2. Telephone conversation with Craig Long, October 1996.

3

GETTING IT DONE

The Role of Senior Managers
Mechanics of a Complete Quality Process
The Baldrige

Quality challenges organizations to turn the ordinary into the extraordinary. It requires senior management teams to convey to all employees that quality is not something extra added onto their "normal" job, but quality *is* the job. This takes some doing.

Designing a quality process that takes into account both the theoretical and the practical factors, that appeals to employees on both a rational and an emotional level, does not require starting from scratch. Happily, there are tools available to help. This section provides some of them, with the best of them being—without serious question—the Malcolm Baldrige Criteria for Excellence.

6

THE ROLE OF SENIOR MANAGERS

While the importance of senior managers to the success of the quality process has been mentioned on several occasions in this book, "A Quality Beginning" serves as a quick review of specific steps to be taken during the definition and initiation of a quality process. If a process is not carefully designed at the outset, the organization will have to expend resources correcting the process after it is launched. As with any aspect of business, it is far easier and far less expensive to prevent problems than to identify and correct them further down the line.

After defining the term "top management," "Top Management Commitment—What's That?" focuses on identifying top management involvement that is active, obvious, and informed.

"Beginning 'Quality Without Limits'" details the specific actions one organization took to prepare for and launch its 100% employee involvement quality process. Unfortunately, it's a story with an unhappy ending. Despite spectacular early successes, the process never reached its full potential. As employees became more active, senior managers reacted by becoming more and more hesitant to share decision-making power. Ultimately, the company was purchased by its chief rival and the process was discontinued.

Then again, even the best of efforts will occasionally slow down as an organization catches its collective breath.

What to do? How can that initial excitement and high level of improvement be revived? "Try Continuous Involvement Improvement" offers a long list of ideas. Quality improvement procedures must themselves be open to improvement.

In short, senior managers must be active in the definition of a quality process as well as in its maintenance and, when necessary, its revitalization and evolution. Anything less will doom the process.

A QUALITY BEGINNING

Quality? Simple. Just do the right things, the right way, the first time and on time. Meet your customers' expectations. But if it is so simple, why have so many quality programs in recent years fallen short of expectations? How should it be?

What too frequently happens is that the management reminds employees about quality, elicits a short-term response from some, and then it is back to business as usual. A successful effort to focus on quality not only galvanizes all employees into action, it also creates a workforce dedicated to perpetuating quality.

What separates the quick-fix (and quick-fade) from the permanent change? The answer does not lie in which precise method the company chooses to use. The day-to-day vehicle must be pro-active, but beyond that there are a wide variety of forms and techniques which have been successful. There are, however, a number of crucial factors that can be looked for and whose presence can be used to predict the success of a nascent quality process.

COMMITMENT AT THE TOP

The major factor in creating a quality process is commitment from the top. This is not just agreement to give *quality a try for a while*, but total commitment based upon an understanding of what quality means to an organization, an appreciation of the competitive edge that quality brings to any enterprise. Make no mistake, quality makes money. Knowing that, and really believing that, is the basis for total commitment.

WILLINGNESS TO CHANGE

Next, there must be a willingness to change. Change is the goal of a successful quality process. Constant change means constant challenge. There will be inevitable changes in the decision-making process as employees learn to look at their jobs critically.

Those managers who view the involvement of employees in solving problems as a breach of authority are a detriment. The company must strive to develop managers who can cope with these changes. This requires an important difference in management style. At all levels of the organization, the goal is for managers to become leaders. The self-confident lead; the insecure can only manage. This aspect of a company's culture may have to be consciously driven by top management in the early stages of a quality process—without it, the effort is doomed.

The quality process must stress, seek and reward changes. It is, quite simply, easier to find 100 different people who can each improve the company's bottom line (or image) by one percent, than it is to find one magician who can, single-handedly, bring about a 100 percent improvement. The big, risky changes do come, but only in an atmosphere in which change is welcome, in which failing in a good effort is acceptable.

INCLUDE EVERYONE

To take full advantage of the concept of incremental change—indeed, to derive maximum benefit from a quality process—everyone must be part of the process. This requires a non-voluntary, formal enrollment of everyone from the CEO to the newest hire.

The minute a quality process is made voluntary, the message sent is: *around here, quality is optional.* This is a message that no company—particularly no American company in today's highly competitive international marketplace—can afford to send to its employees. Too often, only non-management personnel are involved in efforts to improve quality. The message here is less clear, but equally debilitating. Can it be that only non-management personnel have been making mistakes? Or are the company's managers incapable of change?

INVESTING IN QUALITY

To translate these attitudes of commitment, acceptance of change and leadership into action, a quality process requires a substantial investment. Quality is not free! Money is needed for personnel to run the process, for training, for materials and for a program of recognition, gratitude and celebration. The first year's investment is an act of faith based upon easily obtained facts concerning the financial value of quality. After the first year, money spent can be looked at in terms of being a percentage of the previous year's savings and increased revenues due to the quality process.

Investing in staff is necessary There must be at least one person (with appropriate staff) whose only job is quality. Quality is not a part-time job, an addendum to an already busy schedule. The person given this responsibility must have authority to make decisions regarding quality and ready access to the president and/or CEO of the company.

On-going training of employees is a must The opportunity to improve job-related technical skills should be an integral part of the quality process. As a nation, America has built abysmal habits in on-going training. This is due, particularly, to narrow job descriptions traceable in part to union wishes and to a short-sighted lack of understanding on the part of management of the value of thoroughly trained employees.

Invest in recognition Recognition, gratitude and celebration are often overlooked as a key to perpetuating quality. Remember, this is not a bribe to do better; it is a way of saying thank you for a job well done.

These attributes of a successful quality process are not peculiar to any one industry, to service, or to manufacturing. They are applicable in any type of organization blessed with a leadership team which has figured out that quality is their key to the future. American employees must be allowed to join the fight for the success and survival of their companies.

Journal for Quality and Participation, *Vol. 11.1.*

TOP MANAGEMENT COMMITMENT—WHAT'S THAT?

If there's one agreed-upon truth regarding quality, it's this: Success is not possible without top management commitment. Almost as consistent as the call for top management commitment is the bewilderment about what the ubiquitous phrase means. Too often, what passes for TMC should really be called TMP (top management permission) or SMC (some management commitment).

Permission alone won't get an organization within striking distance of its potential. Not only is this a lazy assumption; it's horribly ineffective. Commitment encompasses all elements of permission, not just investing money, material and other people's time. Its power comes from being profoundly personal. More than an occasional speech, a willingness to sign large checks for consultants and to hand out awards at the annual quality celebration, true commitment involves investing oneself and setting an example. This commitment must be active, obvious and informed.

Let's consider those three adjectives individually. Being "active" means more than harassing others with statements such as, "You people need to get better." Top managers must improve whatever it is they themselves do, which is a tricky proposition. To seek self-improvement—which assumes the person isn't already perfect—requires tremendous self-confidence. No wonder some managers hesitate! The best way to judge whether top managers are active is to look at the company's last 10 quality improvement decisions and count how many times top management changed as a result of those decisions. The more decisions that involve changes in top management behavior, the higher the commitment.

There are two ways to define top management. The most obvious one means the people at the very top of the corporate ladder. They are the most senior, their names fall at the apex of the pyramid and appear in gold leaf on their office doors. Their commitment is vital.

A more subtle definition requires looking at an organization from somewhere inside. For most people, top management means anyone three levels higher than the person whose perspective is being considered. If you can make my boss's boss nervous, then you are top management. If you have three or more layers of bureaucracy below you, congratulations, you are part of top management! And that is true no matter how many layers are above you. The commitment of this second group is essential for quality, and it's based on the first group's commitment.

Thus, insisting that every top manager must be "obvious" in his or her commitment is a tall order. When someone in management does contribute to improvement, the word needs to get around. Quite simply, how can anyone else follow their lead if no one is aware of the good example being set?

Lastly, top managers must be informed. They must keep up-to-date in their reading about the quality world, emerging ideas and superlative examples (e.g., Baldrige winners), about successes and challenges within their own organization. They must take part in formal and informal discussions inside and outside the company, demonstrating that they care enough about this quality thing to become intellectually involved. When involved in a discussion where something new comes up, particularly with another member of their organization, they must listen carefully and obviously and ask questions. Being informed is not a state one reaches after a one-time effort; it is a continual challenge.

Which brings us back to the prerequisite for all commitment: self-confidence. Not only is self-confidence necessary to improve one's own work, it's required when a senior manager leaves work behind on her or his desk and gets out there to talk with employees about what they are doing—and sharing what they themselves are doing. It requires self-confidence to let others make decisions about how work should be performed, which is the very essence of empowerment and a keystone of quality. Ceding power and pushing it down to the proper level means developing self-confidence in those lower in the organizational hierarchy so that they, too, can handle the challenges of quality.

But this need for self-confidence should come as no surprise. Only the self-confident can lead; the insecure are doomed to remain managers until they gain the confidence to improve. Quality takes leadership, not management, and top management commitment is one of the distinguishing characteristics of that leadership.

Quality Digest (Internet), *July 1998*

BEGINNING "QUALITY WITHOUT LIMITS"

A quality process in most American organizations is like a fire extinguisher kept in a little glass case down the hall from the offices of the top executives. Printed neatly on the glass is: "Break glass. For emergency use only." All too often, by the time an emergency is fully sensed, either the flames have engulfed the extinguisher or nobody remembers how to use it.

Beginning a quality process, particularly one which formally involves every person on the payroll in a non-crisis setting, is extraordinarily rare. Two examples are the Paul Revere Insurance Group's *Quality Has Value* process (discussed in *The Quality Circles Journal*, September 1985) and the more recent McCormack and Dodge *Quality Without Limits* process. The latter borrowed heavily from the former in its definition stage. The principles are the same; the techniques are specific to McCormack and Dodge (M&D).

In mid-1987, M&D made the decision to establish a formal quality process that went beyond the vital but limited-by-definition software quality-assurance efforts already in place. The timing was based on the belief that the best time for M&D to open its lead on domestic competitors and to prepare itself for anticipated challenges from overseas (most likely from Japan and Germany) was while business was good.

MANAGEMENT SUPPORT OF QUALITY

The groundwork for the Quality Without Limits (QWL) process had already been laid by company president Frank Dodge when he published his basic operating goals, called the *Four Basics*, in 1984. By articulating and circulating his goals for the company, Dodge prepared the M&D employees for a quality orientation. The Four Basics are:

- To provide our customers with the highest quality products that we can.
- To support our customers to ensure their highest possible level of satisfaction with M&D.
- To invest significant resources, financial and personnel, to ensure that our products are and remain abreast of technological advances and remain state of the art.
- To devote significant attention and energy to making M&D the kind of company that attracts and retains the highest quality employees.

His often-repeated statement since the launching of the QWL process that, "If it comes to a choice between quality and schedule, there is no choice...it's quality," has served to strengthen the process. Management support for the process has been demonstrated in a number of other ways. The quality process does more than solicit ideas from employees, it provides a structure for the ideas to be implemented without prior management review. There is management commitment at all levels to actively insure that authority matches responsibility throughout the company.

EVERYONE WILL BE INVOLVED

The key decision in designing the process was to formally involve every person on the M&D payroll—management included—on quality teams. This decision was reached after a simple question was posed: "Whose expertise can we afford to exclude?" The answer was no one's.

TRAINING, TRACKING AND RECOGNITION DESIGN

The primary objectives of the quality team leader training sessions were to give the team leaders background in two major areas: problem solving and running team meetings in a participative manner. The team leaders were, for the most part, the "natural" work unit leaders throughout the company—at all levels. It is anticipated that, in the coming years, the opportunity to be a team leader will be extended to virtually any employee.

A quality tracking system was designed, using a powerful data-management system, M&D *Millennium*. The system has a wide range of data-handling and report capabilities to aid quality team leaders in entering and "stealing" ideas. The company uses the information both to track progress and to know when to say thank you for employee efforts.

The program for recognition, gratitude, and celebration bears a strong resemblance to that established by Pat Townsend at Paul Revere Insurance. When a quality team completes ten quality ideas of any size, or completes a lesser number of ideas with an annualized value of $10,000 (savings or added income), the team is designated a *Bronze Team*. Twenty-five certified ideas, or $25,000 in value brings *Silver* recognition; fifty ideas or $50,000 equals *Gold* recognition.

The actual recognition scheme was designed with one central fact in mind: different folks hear "thank you" differently. At the *Silver* level, for instance, recognition takes place in a number of ways. Team members receive a $25 American Express Gift Check and a silver quality pin from either the president of the company or one of his direct reports, as well as having their name in at least one company publication.

The establishment of a central "quality resource services" department meets the need to have continuity in the day-to-day operations of the process. It also sends a powerful support signal—that through the commitment of resources, both money and personnel—the process is permanent. Heading the department is a senior manager who is a long-time employee of M&D but with no previous experience in quality (specifically in software quality assurance). She has a thorough knowledge of the company and its products, and is highly respected throughout the company. She has four quality idea analysts (two in Natick, one in Dallas, and one in Chicago) and a technical quality manager. The former director of quality resources is available as an adviser.

A YEAR OF QUALITY WITHOUT LIMITS

The outline of the *Quality Without Limits* process was presented to the M&D executive and management committees in October of 1987, and won quick approval. In addition to 100 per cent employee involvement, the plan called for quality team leader training, a sophisticated tracking program and a vigorous program of recognition, gratitude and celebration. Quality team leader training began in Huntsville, Dallas, and Natick in the spring of 1988.

As formal preparations for launching *Quality Without Limits* proceeded, awareness efforts began to take place through a number of channels: the director of quality resource services spoke at the Winter corporate meeting in December of 1987, describing the process-to-be, followed by presentations to anyone who asked him; articles appeared in company publications; and job notices were posted for quality idea analysts.

The next few months were active ones. The Dallas and Huntsville offices finished their quality team leader training in April, enabling their quality teams to get a jump on the official launching of the *Quality Without Limits* process.

By June 7, 1988, the Natick office—and the quality tracking system—were ready to go. To mark the day, the president and all of his direct reports met each employee at the doors of the building, welcoming them to the quality process, and giving each of them copies of three quality posters that had been produced by M&D's creative department.

In July and August, field quality team leaders finished training and the field quality teams were added to the quality tracking system.

Thanks to the training and the repeated, consistent messages that had been broadcast during the previous six months, quality teams and quality team leaders knew that they were being given authority commensurate to their responsibility, and that they were expected to act. In Dallas, within two weeks after teams had been given the green light, a staff member was overheard to say, "Gosh, I sure wish they..." Her companion cut her off with, "Hey, hold it, you can't say that any more. We're 'they'. We'd better fix it ourselves."

After four and one half months, the 147 M&D quality teams had entered 1,828 quality ideas on the quality tracking system, with 530 of them implemented and subsequently certified by one of the quality analysts. While savings are already in the millions, the important indicator of the health of the process is the number of ideas on the system, combined with the fact that their number has grown by a remarkably constant 90 to 100 ideas per week for the last three months.

KEEPING UP WITH QUALITY TEAMS

A process for change must itself react and change. Flexibility is absolutely necessary—along with a willingness, an eagerness, to grow. The process already looks different from when first defined. One of the first problems faced was a happy one—an idea that was "too big."

About McCormack and Dodge:

Founded by Frank Dodge and Jim McCormack in 1969, M&D develops and sells mainframe applications software and software services. Purchased by Dun and Bradstreet in 1983, M&D has long been a profitable organization.

McCormack and Dodge is composed of approximately 1,800 people, split into three general groups: approximately 900 in three central offices (Natick, Huntsville and Dallas); approximately 400 field employees in sales and customer support positions throughout the United States (regional offices in Los Angeles, Chicago, San Francisco, Atlanta, Washington and Newark); and approximately 500 in 15 countries other than the US. (The only continent which is not the home of an M&D office is Antarctica.) For implementation purposes, all US employees would become members of quality teams in 1988; the international staff would be enrolled in1989.

One of the first quality ideas implemented saved the company several hundred thousands in "hard dollars." That was far beyond what was envisioned to become a *Gold Team*; so a *Diamond* level of recognition was invented.

That idea was a mixed blessing. It deserved, and received, special attention. Yet the heart and soul of a quality process, a process aimed at establishing an atmosphere of continuous improvement, is "small" ideas. A conscious effort is made to remind employees that dollar savings is not the focus: quality is. Many implemented ideas will have no particular dollar value and for those that do, the amount is not the overwhelming consideration.

The challenges facing M&D's *Quality Without Limits* process are: ongoing efforts to insure that every quality team is active in the process, expanding the process to the international segment of M&D in 1989, and guiding the evolution of the process as the employees of M&D become more comfortable with the concept of quality.

THE BENEFITS OF TOTAL PARTICIPATION

The *Quality Without Limits* process is an excellent example of the potential of a 100 percent employee involvement process. In making a decision to "begin with everybody," rather than a phased-in program laden with pilot programs and test cases, M&D's leadership reaffirmed its belief in the talents and intentions of all its employees.

The technical software quality assurance efforts of the organization are supported, not replaced, by the QWL process. By involving everyone in the effort to insure that M&D becomes known as the "quality alternative" in its field, the QWL process insures that quality assurance receives more attention than it did when quality was the responsibility of only a few specialists. At the same time, the quality assurance procedures themselves are now subject to improvement—as are all procedures in every department of the company.

McCormack & Dodge, through its implementation of a 100 percent employee involvement and quality process, reinforced with an insistence on pushing authority down to the level of responsibility and thanking employees for their contributions, is not only insuring its own future. It is helping to define an American approach to quality.

Journal for Quality and Participation, *Vol. 11.4., Dec. 1998*

TRY CONTINUOUS INVOLVEMENT IMPROVEMENT

Okay, it's been several months, maybe even a couple of years, and the company's quality process is showing signs of a serious loss of momentum. The exciting big savings racked up in the opening months are institutionalized (and forgotten), and the whole thing is becoming a real chore ... just one more *top* priority item on a long agenda.

Now what? The first step to getting the process back on track is the same as the step necessary for beginning a new process—top management commitment. The men and women at the top of the corporate ladder must agree that there is a problem worth solving and that they will be a part of the solution.

Without the active, informed and obvious commitment of the president or CEO and his or her direct reports, any effort will fall short of its potential. If unable to secure the rational and emotional commitment of this group, it is time for the director of the quality process to quietly update his or her resume... and begin circulating it.

Winning that commitment may mean pointing out the obvious. Quality makes money. Quality is good business. It costs less to do it right the first time. And reliable, quality products and services demand a higher price in a free market. A director intent on revitalizing a quality process should first do his or her homework and, then, armed with incontrovertible proof that this is a serious bottom-line proposal, speak with top management.

There is extensive documentation available to verify that quality is a moneymaker – from data in books by people such as Dr. W. Edwards Deming and Tom Peters to articles in journals of all sorts. Given that the average estimated cost of poor quality (rework, scrap, lost sales, etc.) for American companies is in excess of 30 percent of gross sales, the position that quality has a larger potential impact on the organization's profits than any other proposal on the table is more than just defensible—it is undeniable.

On the other hand, if you think other arguments might win the day when top management isn't interested in getting behind something that can bring in more money, the idea that a quality process is the right thing to do because it improves the lives of the employees (or, even, that it creates or preserves jobs) is unlikely to have any appeal.

SPECIFIC POSITIVE STEPS FOR TOP MANAGEMENT

Because of their position and visibility, managers must take a variety of actions for reasons of both style and substance. Options include:

Quality process as part of staff meetings Begin every staff meeting with a discussion of the progress of the quality process and end each of those meetings with a discussion of the article-of-the-week handed out at the previous meeting.

Publicize staff involvement with quality Once this meeting format (zealously held to) is set in place, it should be thoroughly publicized, both informally and formally. Perhaps the simplest way would be to post the skeleton agenda of top management staff meetings on company bulletin boards.

Model the process Set the example, not by forming yet another task force to figure out ways to tell someone else how to do his or her job better, but by figuring out ways to improve their own jobs. Top management should form its own formal quality team with regular meetings. Again, once management establishes a pattern and has some successes, publicity is called for. Besides the example, the improvements themselves are worth the effort. The quicker, smaller and more numerous the ideas for improvement, the better. Small ideas are preferred because those can be emulated—or beaten. Million dollar ideas are wonderful, and they will appear from time to time in the life of a maturing quality process, but they aren't much of an example. Who believes that everyone can duplicate a million dollar idea?

Learn and use the quality language Make quality part of the staff vocabulary. No member of top management should ever talk to anyone associated with the company for more than five minutes without mentioning quality in general and/or the company quality process in particular. Certain key definitions should be memorized so that every member of top management is sending, at the core, the same message.

While Tom Peters is right when he points out that employees care far more about the movement of the top manager's feet than the movement of their lips, once they have gotten out of their office and moved somewhere, they should say the right things.

No magic ... just thank you will do Get involved with the program of recognition, gratitude, and celebration. If the top management is not already personally, and frequently, saying "thank you" to deserving members of the organization, the failure to do so needs to be addressed.

What's in it for the top executives? Information. (Synonymous in some business dictionaries with power.) When saying thank you to an individual or group, it is only natural to ask, "Gosh what did you folks do to earn this?" Any top manager who wants to know what is happening, what is changing, and who the truly effective junior and middle managers are in the company need only become involved in the effort to take note of improvements in the quality of the company.

Gratitude is extended in a quality process for two reasons: because the recipients deserve it and because if they (the recipients) really hear that the company is grateful, the odds are very good that improvements will continue. Why else should top managers do this? Because it is a great deal of fun.

It's not free, but it has a great ROI Increase the budget. Quality is not quick; it is not easy; and it most definitely is not free. There is, however, no better investment available. Again, this is provable using both external data and successes to date. In short, if a $10 investment has already paid back $20, and the same arrangement is available for a bigger bet, why not put in $100 and double the investment again?

Note, that making a larger budget available to the persons responsible for the day-to-day operations of a quality process (and holding them accountable) is not the same as simply throwing money at a problem in the name of quality. The latter is not recommended.

HIGH HURDLE MANAGEMENT

Besides the involvement of top management, there are two major hurdles encountered by every quality process: middle management resistance and misunderstanding of what a quality process does. Neither of these are necessarily present at the beginning of a quality process, but they are more than likely to develop. And both can slow momentum and impede progress.

Middle management traumas Pay careful attention to middle managers, for their world has undergone the most change in the last decade or so.

Whither the middle? Many quality processes aim all of the initial efforts at the two ends of the corporate ladder, ignoring the middle. While top executives invite non-management personnel to take part in the quality process and urge them to devote a percentage of their time and energy to build the company of tomorrow, these same executives continue to pressure the middle managers to meet the number of today's company.

You can decide when Middle managers are people who have, in many cases, given a dozen or more years to the company and who remember that when they first came to work and presented their first several ideas to their manager, they got a quick rebuff. "Yeah," they were told. "Right. Tell you what, you hang onto that idea and 12 or 15 years from now, if you do it my way, you'll get this job. Then you can do it your way. But, until then, we've always done it that way, and I like it that way... understand?" So they waited and they did it someone else's way. Now they do have the job, and the rules have changed. They are being told to listen to their subordinates for ideas. The logical reaction is, "When the heck do I get my turn? What happened to my chance to make decisions?"

Who, What am I? In a very real sense, they are right. The job they expected to hold has changed, and they did pretty much miss their turn to unilaterally direct operations over the levels they've already climbed above. In the past, what distinguished middle management from non-management was this authority to make decisions. In an effective quality process, however, authority commensurate with responsibility is resident at every level, including non-management. Decision-making is diffused.

New roles and tasks There are, however, tasks that only mid-managers can perform. It means redefining the role of mid-managers away from day-to-day micro-managing toward a role as a trainer and communicator. In addition to continuing to make decisions appropriate to his or her responsibilities, the middle manager must help his or her subordinates to better understand and maximize their contributions to the company. Rather than simply being the carrier and overseer of orders from on high, these managers now take on the role of passing information both ways. It is on their turf that the top-down commitment and the bottom-up implementation integral to a successful quality process comes together. While these responsibilities are ultimately more satisfying, they also constitute a more difficult and a more complex job; one that was not expected.

Turning a hurdle into a launch pad There are two key ways to turn middle managers into active partners in a quality process.

Involve them Middle managers should be included in defining the evolution of the quality process itself.

Give them a power transfusion The company should try to insure that these managers are receiving power from higher up the corporate ladder even as they are transferring some of their old power down; it's as simple as actively soliciting their ideas in the same way that they are expected to solicit ideas from their subordinates.

Don't underestimate this problem. The support of these managers is invaluable. To ignore them is to invite everything from damning indifference to outright sabotage.

The quality time hurdle The second major hurdle is the idea that quality does not require a pro-active process. It can best be described as the idea that quality isn't special, "We do quality work all the time." Some managers, uncomfortable with the idea of the loss of authority that is requisite, and natural, to a 100 percent involvement quality process, will argue that separate meetings aren't needed and that they will simply "include quality in my regular staff meetings."

Resist that idea!

Quality is special, and it needs a separate, discrete time devoted to it. Almost all meetings held in American businesses center upon discussions of how to catch up or, at best, how to solve today's problems. Quality meetings are focused on tomorrow, on changing systems from *good enough to world class.*

In a regular staff meeting, if there are 15 minutes scheduled for today's crisis and 15 minutes for tomorrow's possibilities, it is virtually inevitable that 29 minutes will be spent on today's crisis and one minute will be spent saying, "We'll discuss this quality stuff next week, so hang onto those quality ideas. Thanks for your help in getting this fire put out."

If American businesses really did do quality work all the time, the reputation of American goods and services would be better; it might even be that the country's trade balance would be leaning in the other direction. Any top manager who truly believes his or her managers when they say that "it isn't necessary to pay special attention to quality," should check their company's latest market share or profit figures. Those with solid profits and growing market share know that quality takes a single-minded, specific effort. Those with less secure futures probably got that way by assuming *we do quality work all the time.*

BEYOND HURDLES TO RENEWED VISIONS OF QUALITY

Once the company has a clearer vision of why a quality process is indispensable, there are a numerous options for revitalizing the process:

Involve everybody If the quality process to date has been a volunteers-only approach, or some other approach that results in less than 100 percent formal involvement—it is time to declare that effort a successful pilot program and get on with the real thing.

Form a new top quality committee With the support and understanding of top management, trumpet the successes that have been achieved, and keep it going while building a newly formed quality steering committee that should include members of top management (a good guideline: the number one or number two person from each division or major department) and the leadership of any existing quality programs.

Just the establishment of a committee consisting of senior vice presidents and some of the coordinators, trainers, leaders, and facilitators (usually mid-managers) who have kept the current program going will send a strong signal. Few people have ever heard of such a mixed-rank committee, much less been affected by one. This committee should have the freedom to consider a major overhaul of the current process, but should also have the sense to take full advantage of the people and experience at hand.

Give real authority One major point to reiterate: the level of trust inherent in an effective process. For maximum impact and return on investment, quality teams must be given the authority to make changes within their areas of responsibility.

Any group of people required to ask permission to do something that they know is right, and that will impact only their work, fully realizes that they are not trusted. Understandably, their motivation to look for a way to improve procedures will lag far behind that of contemporaries in other organizations that treat their employees as adults.

Learn how to say Thank you! Update the program for recognition, gratitude, and celebration. Two rules of thumb: say "thank you" in three or four different ways and make it enjoyable.

An example of the first: when the Paul Revere's *Quality Has Value* process was in its second year (it is now in its sixth year), the thank you that was extended to each member of a team at the silver level (25 completed ideas or $25,000 in annualized savings) consisted of:

- Two $10 gift certificates (redeemable in approximately 50 stores and restaurants in Worcester, Massachusetts)
- A Silver quality pin or quality charm presented during a ceremony conducted by the president of the company or one of the two co-chairpersons of the quality Steering Committee.
- Recognition in the *Quality News* (a monthly news letter).

By using a variety of *thank yous*, a company increases the chances of each employee's actually hearing the message.

An example of the second: when the McCormack and Dodge *Quality Without Limits* process was about four months old in 1988, National Quality Month (October) was noted by having the members of the executive committee visit quality team meetings with bags of Halloween candy.

There has to be a core program of gratitude and recognition so that everyone knows with certainty that they will be thanked for progress. In addition, there should be events designed simply to celebrate and bolster unity and morale.

"Turbo" quality Launch a supplemental, singular, high-visibility, everybody-can-do-it, short-term quality program. The program chosen must be consistent with the overall goals of the quality process and should be conducted in an atmosphere of fun. For a service organization, a Phone Perfect effort for a designated length of time is one option.

For a number of days, set as a goal that no telephone in the company will ring more than three times and that every phone will be answered cheerfully and politely. Have someone (a Phone Phantom) make random calls and reward the person who answers—if they do it correctly (a pocket address book would be appropriate). Set up scoreboards in departments on which people can keep track of how many phones were answered on the first, second, third, or beyond ring. Have executives roaming the floor, congratualting departments with good records. Put the CEO on the switchboard for a while. When it is over, celebrate the success.

And don't be heavy-handed with the "Why can't we do it this well all the time?" messages. You've got adults on the staff; they'll see what is possible.

Tell people about the good news Publicize the efforts. Use all the media available (a monthly newsletter is a good start) to publicize the successes. Tell the other employees; tell the business media; tell your paying customers. A monthly newsletter should be well-written and, every month, include the names of the most recently successful teams and people, a piece of news about quality from the outside world, the description of an achievement of note within the company, and a philosophy piece. If there is room, include copies of a letter or two from satisfied customers (internal and external) to company employees.

Add new links to the quality chain If, for instance, the Quality Steering Committee (either newly formed or newly interested) feels that training in participative management is appropriate, make sure that everybody knows that the reason the new training program is being put in place is to increase the support for the quality process.

If a survey (formal or informal) reveals that no one knows what the company policy statement is, have it rewritten to include quality—and involve a lot of people at various levels in the redrafting effort. Then have it printed up on little cards, add the quality process logo, and give one to every employee. In other words, make attention to quality (as well as recognition of the existence and importance of the quality process) all-pervasive, from the CEO to the newest hire.

Some of the ideas listed above can be tried in a top management vacuum to help a lagging process. However, the success of most of them depends on the cooperation, if not active participation, from upper echelons of the company. It all comes back to top management commitment.

Whether getting started or picking up speed, the full potential of a quality process will never be reached until top managers can get on an elevator, notice the quality pin of the non-management person next to them and feel comfortable saying: "My team is finally going to earn our quality pins this month. How did you get yours?"

Journal for Quality and Participation, *June 1989.*

7

MECHANICS OF A COMPLETE QUALITY PROCESS

"What Went Wrong with Quality"? begins where the previous chapter left off—by discussing things that can go wrong with a quality effort if senior managers fail to show sufficient leadership or support. On a more positive note, it goes on to list the seven elements of a Complete Quality Process. These seven elements are to be reiterated in greater detail in the last article in this chapter.

Every process is unique in its final details, but they all share some basic principles and characteristics as they evolve. "Four Phases of a Quality Process" examines evolutionary characteristics that virtually every individual and every section or division of an organization will experience. Knowing what to expect helps everyone understand what is going on around them and enables them to react in productive ways.

The last article in this chapter—"Quality Is Everybody's Business"—is the transcript of a speech given by Patrick Townsend to an audience of American newspaper executives. While introducing both new ideas and new examples, it also summarizes much of what has been covered to this point of this book.

WHAT WENT WRONG WITH QUALITY?

Nothing has gone wrong with the concept of quality; plenty has gone wrong with the implementation. Much of the trouble can be traced to a failure on the part of senior executives to recognize the link between quality and leadership.

Take as a definition of leadership what was proposed earlier:

Leadership is the creation of an environment in which others can self-actualize in the process of completing the task.

Now look at any all-too-typical company (or government organization), especially any service company which currently houses quality cynics and critics. Chances are you'll find that the company turned over millions of dollars to a "quality consulting" firm in the late 1980s or early 1990s.

Most of the major consultant companies simply did not do the intellectual heavy lifting necessary to devise an approach to quality consistent with the peculiar demands of the American workforce in general and the challenges of service organizations in particular. The scheme proposed by the majority of leading companies (and most of the small ones as well) went something like this:

1. The CEO/President was personally convinced by the head of the consulting firm that pursuing quality was vital and that enlisting the aid of a large, expensive consulting firm was imperative due to the complexity of the challenge.
2. The CEO/President attended a couple of lectures and made a few grand speeches and then effectively disappeared – reappearing once a year to make a statement concerning his or her continuing support for the "quality program."

3. Senior executives of the organization were formed into a "Quality Council" (or some similar name) that rubber-stamped the consultants' plans for training and for the structure of the quality process.
4. Virtually everyone on the payroll was exposed to a form of "quality training." This evolution took place over a couple of years due to the consultants' insistence that everyone be trained in problem-solving techniques, etc.—usually before expecting bottom-line results. The quality consultants' battle cry, "This will take a long time and cost a lot of money" was self-serving and wrong. Looking for perfection as defined by a team of consultants and backed up by many long months of expensive training and preparation bred cynicism. It took so long to see any results that those results were bound to look paltry compared to the investment in effort and money.
5. Quality processes themselves were devised along one of two lines:

   In the first, top layers of councils (or "groups" or whatever) set policy for, oversaw the activities of, and approved the recommendations of the next lower level. At one major organization in the Northeast, for instance, one of the leading consultant firms defined a structure in which there were literally five layers of councils overseeing the next lowest level. Only the 6th level down (exclusively non-management personnel) was actually charged with making improvements!

   Alternatively, a process focused on only one subset or one component of quality (as with re-engineering, see below). These unbalanced efforts saved the senior executives from having to trust lower-level employees with any meaningful decision-making.
6. With a fair bit of fanfare, the process was launched—and then ignored and allowed to wither. In the end,

a small group of quality groupies made just enough noise so that the organization could mark the "yes" box when the surveys came around asking, "Does your organization have an active quality process?"

Contrast this with the definition of leadership. The deviation from the basics of an environment in which others can self-actualize within the framework of their job responsibilities often began when the titular head of the organization disengaged and left the definition of the environment of the organization to someone whose primary goal was the existence of his or her own consulting organization. It continued when consultants instituted some sort of warmed-over version of American Quality Circles with an elaborate system of multi-layered over-the-shoulder supervision calling for a series of approvals from higher levels before an improvement could be implemented. There is a better way.

With or without a consultant, an effective approach to quality includes several essential components:

Top management commitment This must be active, obvious and informed. It can be best demonstrated through a personal commitment of time and ego, backed by organizational resources. One option is through involvement in the committee that defines the mechanics of the organization's quality process; another is through involvement in the organization's program of recognition, gratitude and celebration; still another is through active and ongoing efforts on the part of the senior executives to improve their own work procedures. As one quality entrepreneur in Texas puts it, "Their hips and their lips have got to go in the same direction." Ask this question: "Of the last 10 decisions made by senior management 'in the name of quality,' how many required that senior management change its own behavior?" If the percentage is low, there's trouble ahead.

Leadership There must be a common understanding throughout the organization of what "leadership" means.

This does not happen without frank discussion, training, and empowerment (authority equal to responsibility). Encouraging employees throughout the organization to practice leadership behavior is the hallmark of an active quality process. This requires a leap of faith; a hop of hope won't cut it. (Various aspects of leadership will be explored in this space in the months to come.)

100% Employee involvement – with a structure It is logically indefensible to involve less than 100% of the people on the payroll in quality improvement. Whose help don't you want and need? It is unrealistic, however, to ask people to leave their comfort zone and embrace change without spelling out clear guidelines for how things will get done in the future. These guidelines must include ways to address both the question, "Are we doing the right things?" (re-engineering, value analysis, and blueprinting are three alternatives) and the question, "Are we doing things right?" (quality teams in which every person on the payroll is on at least one team or a Milliken Industries-style suggestion system are among the alternatives). Ask this question: "Where do you go with a good idea?" When every employee at every level of the organization knows the answer and the answer encourages active participation, you're getting somewhere.

Communications Remember two things: What counts is what is heard, not what is said; and upward communications is as important as downward communications. Structure procedures to help all employees improve their personal communications style. (Practical ways to improve communications will be the subject of a future column.)

Training Instead of waiting for every person to go through quality training, an organization is much better advised to train 10–15% of its employees as "quality team leaders" who can lead groups of employees through problem-solving discussions. Other employees can learn by observation or, in future years, through formal training when it is their turn to be a quality team leader. Keep problem-solving simple. A major fault with quality training (besides the inflated prices) has been the requirement to drag every single problem through every single step of the consultant company's seven- (or nine- or eleven-) step process. Many problems can be solved effectively by saying, "We do WHAT?!? Good heavens, let's not do that!".

Measurement Misuse of measurement ranks high as a source of trouble. While vital to success, measurement is not a religion and should not be used as such. There are no mysteries here. Benchmarking may sound complicated but everyone has heard of the Evil Queen who asked, "Mirror, mirror on the wall, who's the fairest of them all?" She was benchmarking – and to give her credit, when she got data, she acted. Employees rightfully resent measurement when it is used as a source or form of punishment. There are only two legitimate reasons to take measurements: to track progress and to gather data that can be a source of ideas for improvement. Individuals charged with collecting data must be able to answer the questions, "Why are you taking that measurement? How will the data be used?"

Recognition, Gratitude & Celebration An organization must say thank you. Difficulties often arise because different people "hear" thank you in different ways. (You will find this topic fully covered in *Recognition, Gratitude & Celebration* [Crisp, 1997]). Suffice it for now to say that this effort will require a significant investment of time on the part of senior executives. Interestingly, the less time invested, the more money required to get the same results.

The elapsed time between the day the senior management team of an organization of 3000 or fewer people makes the decision to "do quality" and the day when a complete quality process with the active involvement of 100% of the employees is in place and functioning should be no more than six to eight months. (A description of a process that fulfills all of these criteria is the heart of *Commit to Quality* [Wiley, 1986].)

Any process will evolve and improve. Taking a look at your own quality effort and comparing it against the criteria above can give you an insight into what you can do to revitalize your own quality process.

Quality Digest (Internet), *Nov. 1997*

FOUR PHASES OF A QUALITY PROCESS

Like the phases of the moon, a quality process has identifiable and sequential phases of its own. The first three of these were articulated by Jeff Pym, who led and observed the implementation of a successful quality process in one division of a large company. The fourth phase was added three years later by Joe McConville while administering the process from the company's main office.

Quality processes evolve through definite steps:

- What can you do for me?
- What can I do for me?
- What can I do for you?
- What can we do together?

This sequence holds true for both individuals and groups, although the speed at which they progress through the four phases varies considerably. Some people, prepared to trust an organization's senior management, can begin at the third or fourth phase. Others, perhaps more cautious from their experiences with previous programs, may remain in phase one for months on end.

No matter how well-defined a quality process is, or how deeply committed a senior management team, organizations are still made up of individuals who must each make personal commitments if the quality effort is to succeed. Senior managers should know that, while they cannot expect everyone to react at the same speed, they can expect the same sequence of acceptance. This knowledge encourages them to applaud those taking the lead rather than fussing at those proceeding more cautiously.

As a rule of thumb, the length of time that people stay in the first phase— *What can you do for me?*– is inversely proportional to their emotional (and, often, geographical) distance from the organization's hub. Put more bluntly, the announcement of a new initiative by the home office is not always greeted with rampant enthusiasm by troops in the field who, according to collective memory, have seen a lot of programs come and go. No one should be surprised when a salesperson located 1,500 miles from headquarters says, "Until someone actually does something which removes a procedure that makes life hard for me, you can count me as a spectator." By the same token, depending on the organization's past history, the psychological distance between the first and fourth floors of a headquarters building may be every bit as great as the distance from one coast to the other.

Initially, the first phase doesn't seem to fit very well into a corporate agenda. In order to accept a new program or process, an organization's members must be able to see something in the new procedures that will benefit them personally. If the answer to "*What can you do for me*?" is "Nothing," the next questions in the sequence won't even come up.

Fortunately, in most workplaces there are optimists who see quality as a way to control their day-to-day activities. "Hey, someone actually used quality to eliminate that stupid procedure," is a powerful inducement for first-phasers to venture deeper into the process.

The second phase—*What can I do for me?*—while still self-centered, represents the beginning of activism. It's the realization that they can effect changes and share in the decision making that convinces people to become quality enthusiasts.

When the workload is a continual challenge to keep up while staving off problems, no one has the time or inclination to help others fight their battles. But once personal distractions are removed, individuals and teams can begin to reach beyond their immediate borders. Then the move to the third phase—*What can I do for you?*—is an incremental and logical one.

This phase is long-lasting. There is a seemingly neverending list of things that people want in the way of service. Yet, in the interaction between customer and supplier lie the seeds of phase four, *What can we do together?* At the company where this sequence was first articulated, the fourth phase emerged in 1988 during a Baldrige application process. More than 75 percent of the organization's quality teams had communication and training goals, and were looking for ways to partner with other teams. That same year, partnering gained national attention when Motorola won one of the first Baldrige awards.

Once the majority of the people and units within an organization reach the fourth phase, a company's quality process can be considered mature. Caution: "Mature" and "frozen in place" are not synonymous. What we can do together today is different from what we could do together yesterday and what, if we act together today, we can do tomorrow.

Quality Digest, *March 1998*

QUALITY IS EVERYBODY'S BUSINESS

Mr. Finneman: Good morning. As moderator of this sixth annual J. Montgomery Curtis Memorial Seminar, I'm very pleased that you could all find the time to be with us. I'd like to emphasize one thing before we get started.

During this program, we want to hear from everybody. We are looking for the best ideas from each seminar member, as well as from our four scheduled speakers.

Each of our speakers has a prepared presentation. But each speaker is also leaving time so members can ask questions, make comments, disagree, whatever the case may be.

And with that, we're going to get under way. I'd like to introduce Pat Townsend to you. Pat is co-author of "Commit to Quality," and president of Townsend and Gebhardt, Advisers on Quality. He's a former Marine turned quality adviser, and has had more than 150 articles published in various periodicals on a wide variety of topics. He also has contributed chapters to two books, one of them "Quality Dynamics for the Service Industry."

Pat.

Mr.Townsend: Thank you. I know it's traditional to start with a joke, but I want to start instead with a quote by a gentleman named Konosuke Matsushita. He is executive adviser for Matsushita Electric Industrial Company, which owns Panasonic. Speaking to an American audience, he said:

"We will win and you will lose. You cannot do anything about it because your failure is an internal disease. You firmly believe that sound management means executives on one side and workers on the other. On one side, men who think; on the other side, men who only work. For you, management is the art of smoothly transferring the executives' ideas to the workers' hands.

"We, in Japan, are past that stage. We are aware that business has become terribly complex. Survival is very uncertain in an environment filled with the unexpected and complications. Therefore, a company must have the commitment of the minds of all of its employees to survive. For us, management is the intellectual commitment by the entire work force, without self-imposed functional or plastic barriers."

When you're talking about quality in America, there's the classic good news and bad news. The good news is, there's growing evidence that this quality, customer service thing is very real, very profitable, and very much a key to survival. The bad news is that the evidence is being ignored for the most part; very few companies are doing anything about it beyond lip service.

For the American press, there are three different levels of possible concern.

One is for your own survival in your own marketplaces—for your own financial benefit.

On the second level, the thing that makes it difficult for you is that newspaper organizations are a microcosm of American society across the board. Within one organization, you've got the straight manufacturing side cranking out a paper that is mechanically good, printed correctly with all the words in the right sequence and spelled right. You've also got—through the use of ideas and information and the servicing of customers—a classic service-type operation going on inside your newspapers.

The third thing I'd like to point out is, you have one additional responsibility. Because of the nature of your business, it's very much your responsibility to tell everybody else about quality. It is your role to inform the American reading public, the American business community about the progress of quality, to give examples good and bad. Quality is good news. It's good news in the sense that it's valid news. It also serves the role of informing and helping the entire American economy to move forward, to know what's going on.

Quality and customer service are intermarried concepts. You cannot do one without the other. Before I start throwing the word "quality" around, let me stop and define it because we're going to have to agree on a definition. Before you begin talking to your employees about quality, you have to make sure you understand what the word means.

That's not an easy task. There was a survey conducted by the American Society for Quality Control. They asked 615 executives—big companies and small companies of all types—what quality meant to them and what they were doing about it.

The answers ranged from frightening to inconsistent. American executives don't have an agreement on this thing. As an example, just review a couple of questions from the survey.

One, how often does your company use the following programs or methods? And it gave them a whole list of possible quality-type methods and processes. One I dearly loved concerned total quality control. Does your company do total quality control? About 38 percent said, yes, we do that very often; 30 percent said we do that often; 16 percent said we do total quality control sometimes; 6 percent said rarely; 4 percent said never; and 6 percent didn't know, which is, perhaps, the most frightening of all those numbers.

But stop and think about that. You've got 30 percent saying they do total quality control often. How in the hell do you do that? Let me rephrase the question. Let's say that you were asked if you were totally faithful to your spouse. And in reply 30 percent said often, 16 percent said sometimes absolutely faithful. I mean it doesn't make any sense; there is no depth of understanding here.

Another of the survey questions is also worth reviewing here. Poor quality—as measured by repair, rework, lost sales, and so on—is said to cost American business billions of dollars annually. They asked those 615 executives, "How much does it cost your company? How much does poor quality cost your company in percentage of gross sales?"

Forty-four percent said less than 5 percent; 19 percent said between 5 and 10; 6 percent said it was in the teens; 3 percent put it in the 20s; 1 percent said 30 to 49, and 27 percent said they didn't know.

The best research in this field says that the actual average for manufacturing companies is in the low 30s, for service organizations in the mid to high 30s. Thirty-five percent of gross sales of the average American service company is being thrown away, is being wasted.

That's why this quality thing is worth talking about. Quality makes money. That's why you do quality. It makes money.

So what is it? Well, definitions abound. Dr. Juran prefers to call it "fitness for use." Phil Crosby says quality is what "conforms to the requirements of management."

A fellow named Bill Scherkenbach, one of the prime movers in Ford's quality process, defines quality as "when your customers brag about you."

And, of course, there's also the old definition that goes with pornography: "I know it when I see it."

Let me offer a definition that I think is practical and that you can use. "Quality exists in two pieces. There is quality in fact and quality in perception. And you have to have them both to succeed."

Quality in fact means you do it the way you think you should. You meet your own specifications. You conform to the requirements of management. You do it the way you've been told to. That's quality in fact—but it's not enough by itself. There is also quality in perception.

One quick example is, of course, the Coca-Cola Company. They made new Coke several years ago. They made new Coke exactly the way they wanted to make new Coke. It was precisely what management had required of the employees of Coca-Cola. The problem was trying to find somebody who would voluntarily drink it. It didn't meet anybody's expectations. It wasn't what Coke was supposed to taste like.

Quality in perception is when somebody else believes that the product or service you're offering is going to do or be what they want it to do or be. Meeting somebody else's expectations–that's quality in perception.

There's another word to define, then we'll tie it together. That other word, which is central to our discussion today, is "customer."

Let me give you a definition: "A customer is anyone to whom you provide product, service or information. Anyone."

Now, the reason I define it that way is because if you limit your definition to just the consumer buying a newspaper, most of the people in your organization will have a tough time with that because they don't deal eyeball to eyeball with that person. It's a neat concept, but it's not practical.

But if I say it's anyone to whom I provide product, service or information, now it becomes practical. I can turn to the person next to me and say, "Is this what you want, what are your expectations? What is it you need from me?"

One of the basic starting places of any quality or customer service process is to figure out who your customers really are. To whom do we provide product, service or information? Make a list, and you'll be amazed at how long the list becomes.

You will also find that two people are often each other's customers, depending on what they're doing at a particular moment. You may be my customer at this hour and the next hour I'm yours.

Let me give you an example. I helped establish and run the quality process for the Paul Revere Insurance Group for a few years. It's a process that has had a great deal of success, I'm happy to say.

In talking to employees, we found out one young lady had just discovered something. She had been there almost two years.

When she first came to work, she had learned her job like many folks learn their jobs—from the person she replaced. Well, they had a three-hour overlap. The new employee was told to do this, don't forget to do that—you know, the whole drill you go through in that three hours. The new employee was told, "Oh yeah, every Tuesday at 1 o'clock this box of cards is going to show up on your desk. Stop everything and put them in alphabetical order, and then just put them on that desk behind you because she works on them next."

Two years deep into this drill, she put the box back there one Tuesday afternoon, turned back to her desk, then looked back again just in time to see the young lady behind her dumping the box into a trash can. Now, she reacted like any mature business person would react: "What the hell are you doing back there?" The response: "Well, I don't use these any more. I started getting a printout 18 months ago."

The fault, if there is one, is with this first lady for not asking what the expectations were. She hadn't talked to her customer. She hadn't said to her simple things like, "Do you use this, is this any good to you?" Those simple conversations are the beginning of a customer service program, of a quality process.

The next time you fill out that monthly, or weekly, report that goes to 82 people or five people, I've got a suggestion for you. Don't put it into the distribution system; hand-deliver it to each person and ask each. "What part of this do you use?"

You're going to find there are chunks that nobody uses. In this case I just gave you, this young lady had not maliciously wasted the company's time for an hour and 15 minutes every Tuesday. She was doing exactly what she thought she should do. What she hadn't done was talk to her customer.

Take that idea of focusing on customers. Who is your customer or customers? How do you balance the needs and expectations of all of your customers? That has to become part of the natural vocabulary of your organization.

It becomes part of the natural vocabulary because you set the example. It becomes part of YOUR vocabulary. It becomes part of what you train and teach people to do.

The whole key then is to determine what your specifications are and what your expectations are, to see if they match. If they don't match, talk about it. Change one or the other, but no surprises.

The trick is be pro-active about it. Don't be passive. Get out there and talk to folks, have folks talk to each other. Communications is the key to a good, solid, long-lasting, quality customer-service process. It becomes a major characteristic of an organization as the process takes hold and becomes part of the corporate culture.

Communications must be consciously attacked and become a conscious part of what you're doing. It's hard work. We're talking about changing habits, changing a corporate culture; that's never easy.

Then why bother? As I mentioned, it makes money. It creates and it preserves jobs. Isn't that simple? It creates and preserves jobs.

It does require, in a sense, a leap of faith, because you've got to invest up front. But it's no more a leap than in any investment you make, because it can be based on a good deal of research and hard facts.

Just look at the evidence. In the American economy, people will pay extra for a product or service if it's actually going to work. In another of those surveys by the ASQC, folks were asked: "If you're going to buy something, how much extra would you pay if I guarantee it's actually going to work?" Eighty-two percent said they would pay extra, by an average of 20 percent, if they felt sure that what they bought would meet their expectations. We don't believe things work in this country unless they cost extra. That's why quality makes money. It takes advantage of that.

Ford is a marvelous example. In 1980, Ford lost $1.9 billion. They figured out they had a problem. About that same time, NBC ran a TV program called "If Japan Can Do It, Why Can't We?" featuring Dr. W. Edwards Deming.

Chairman Donald Petersen of Ford invited Dr. Deming in. Deming gave them a list of things to do to make things better. There was a tremendous effort involved, and tremendous investment. But, as you all know, for the last three years Ford has made more money than GM. After 62 years of General Motors making more money than Ford, for the last three years Ford has made more money than GM on fewer sales.

Motorola was one of the three winners of Malcolm Baldrige National Quality Awards last year. The company estimates that its quality process has been worth $250 million. Motorola came out a few months ago with a cellular phone about the size of a deck of cards. They say it's 18 months ahead of the competition. Competition, of course, meaning Japan.

That's the kind of thing that happens to a quality organization. Good things happen, like being 18 months ahead of your competition.

On the service side, there's an outfit in Canada, AVCO Financial Services, money lenders. They began a quality process in mid-1985. So 1986 was their first full year of having a quality process in place. They began the year as the No. 2 money lender in Canada, but by the end of the year they were No. 1. By the end of the following year, their profits were doubled. By the end of the third year, they had buried their competition. And the only thing they did was a quality process.

There are all kinds of examples out there that justify this leap of faith and justify this investment. But a quality process is not easy, it is not quick, and it is not free.

The first attempt at quality programs in America began in the wake of World War II, almost solely in the manufacturing world. They named quality control specialists, just a small group of people. They gave them a bagful of measuring tools and a little bit of training, and then sat them next to the loading dock and said, "Stop the bad stuff."

Now, think about this for a second. If it's their job to stop the bad stuff, what's our job? Get it past them. Right? Our job is to make bad stuff and fool them, and get it out there on the street. That's the way these companies worked. Quality control doesn't get it, if that's all you have.

My wife and I were in Russia last year, and we found a store where they sold kitchen stuff. While we were looking around, I heard ping, ping, ping, clunk, ping, ping. There was this clerk, and she had a bunch of cups lined up that this customer apparently was going to buy. She took out a spoon and hit each cup. She picked up the clunker, threw it in the trash, got another one out, pinged it and then sold it.

That's the most expensive kind of quality control you could find, because of all the expense of making that clunker and now it's in the trash can. That's what quality control does for you. It catches the bad stuff. That's not good enough. It costs too much to do.

The second round of American quality programs began with a different question. It began with the idea of: "I wonder who we can get to volunteer to be part of this quality thing?" In the mid-70s it began, called "quality circles" with all types of variations of that. You would get all of your employees together and you'd say. "Who here wants to work on quality?" About 10 per cent of the hands would go up. That ignores, of course, that you also were asking. "Who doesn't want to?" Ninety percent of those hands were going up.

You're right back to the folks whose job it was to make things better – and nobody else had to worry about it. Of course, you only had non-management people in quality circles. What does that say to everybody?

Not only that, but once these non-management folks had an idea they wanted to propose, they had to go play "Mother may I" with management. Now, management didn't understand the question anyway. The non-managers had to go in and try to explain what it was they wanted to change, so the managers could say yes or no. It was just all convoluted, and it took a long time.

There were some great success stories at first, but virtually all quality-circle programs died within a couple of years.

The third approach, which is still not used as much as I would like, begins with a very different question. Rather than trying to find what small number of people to involve, look at it from the other side. Who do you want to leave out? Who should we exclude from this quality process?

Now, if you can think of somebody in your organization who you don't want to improve, if you can think of people who you believe are incapable of a creative or interesting thought, that's fine to exclude them—but only if you're going to look them square in the eye and tell them. "Well, Fred or Freida, we at top management have given it a lot of thought. We've decided that you're about as dumb as a rock and couldn't possibly contribute anything to this company ever, so you're not going to be involved in the quality process." If you're willing to do that, fine; otherwise, go for 100 percent employee involvement—active, real, formal enrollment in the process.

Now, what I'm NOT talking about is the annual event where the CEO trundles out of his or her office and says, "This is a quality organization, guys," and then trundles off again with a cheery "See you next year." Perhaps this is bolstered by a willingness to sign the occasional check, or hire the occasional consultant. But that doesn't get it, folks.

Quality and customer service require your personal active involvement. If you're not willing to do that, then stop this thing right now. It's going to require your personal time and your personal effort. You need to be willing to trust the folks on your payroll. You need to be willing to push authority down the chain, all of the authority that kind of gravitated to the top over the years before you got there.

The nature of authority is to move up the chain of command. How does that happen? Easy. I'm doing job A and I'm promoted to A+1. But I know how to do job A, so I take that with me up to A+1. And I keep dragging stuff with me as I climb the ladder; all of this tar goes up with me. What I leave behind, the next people bring up with them when they get promoted.

The power has drifted upward over the years. You've got to push it back down to where it belongs. To trust the folks there will require a great deal of self-confidence on your part. But to me, that's only fair and and that's only logical, because only the self-confident become leaders. The quality process needs leadership, not management.

Too many American companies' top management can be described in this way. They get up in the morning. They get dressed, get in the car, drive to work. All the way to work, they use their brakes. They have never met the person who put that brake assembly together.

But they use those brakes again, again, and again. They go through all kinds of intersections, never even wondering if that other driver is going to stop at the red light.

They get to their building, park in their reserved space, go in and push the buttons to their floor. They never look to see who signed that little certificate that says the elevator won't fall into the basement today.

They finally get to their office. By the time they sit down at their desk, if you review their previous half-hour or so, they have trusted their life to dozens, if not hundreds, of strangers.

Now comes an opportunity to trust Mary Jane with a $25 decision. She's been here for 30 years, and the manager says no. It doesn't make any sense. We'll trust our lives to strangers, but won't trust Mary Jane with 25 bucks. It's illogical to treat your people like that.

Okay. So what should you do? Let me go through a set of principles. These principles cross virtually all boundaries between organizations, between countries, anybody you want to name.

The techniques to adapt those principles to your particular organization, however, are your problem. Each organization comes to the table with a different set of habits, a different culture, different problems. So the techniques have to be adapted. Principles, however, can be adopted wholesale.

The first principle of course, is top management commitment. What I mean by that is your activity, your involvement. You and the other top management folks of your organization must be informed, you must be active and you must be obvious. The involvement must be all three: informed, active and obvious.

To be informed, you have to just read. There are lots of books, lots of articles, case studies and examples.

Once you are informed, you become active and obvious. To do that you've got to get out of your office. Go out and talk to real people, real employees and real consumers. People have to see you doing this, otherwise it's going to be one more program of the month: "Yeah, yeah, this too will pass. We'll just hunker down here and let it go."

Informed, active and obvious. You're going to have to give your company's time and your company's money—but you also have to give your personal time and energy, because what else do you have to offer?

The folks on the payroll care about who's on top. They watch you. And if they don't understand you, if they misread you, then whose fault is that? It's yours. You're the one who is sending out the messages. They care who you are. They care what you do. They will emulate what you do.

You need to understand that leadership is something that can be learned. It can be taught and it can be learned. It's not something you're born with necessarily. Now, I grant you, great leaders, special leaders, are born with something extra. Einstein was born with something extra—but we've trained lots of good mathematicians.

I suggest that you investigate the military as a possible model of how leadership can be learned. I grant you that there have been, and still are, enough idiots in uniform to keep a negative stereotype alive. But they are five percent.

Military leadership theory has evolved from an arena in which the penalty for error was very quick and very severe. The US Army's manual defines three styles of leadership: authoritarian, participative and delegative.

Authoritarian will only work if used sparingly and these three conditions prevail: You are out of time, you have all the information you need, and your people have high morale. They have high morale because you don't use authoritarian leadership very often. You normally use participative and delegative.

Under the participative style, you look for input before you make the decision. You have the authority and the responsibility, but you get the information. You ask, "What do you think? What do you know? Give me your information." Then you make the call.

Delegative is when you push the authority down below the responsibility. The responsibility is still yours, but you delegate the authority: "I'll cover you. I'll back you. I'll support you."

Quality is what I consider a very basic thing. Every single person on the payroll becomes formally involved in the process. How would I do that? I break them into teams.

Every person is enrolled on a quality team or whatever name you give these little groups of people. That involves training, some organization, a tracking system. It involves pushing authority down to these teams and saying: "You get an idea, if it falls within your area of responsibility, do it. Make it better, now."

Please notice, I'm not urging that you turn them loose. I'm not suggesting a do-your-own-thing kind of place. No, this is an ordered business process. It's authority equal to responsibility. Folks in the mailroom can't change a vacation policy, that's not their responsibility; but they can surely change the mail delivery route.

Once you give that kind of authority, you will find that your folks become tightly focused on their jobs because they can actually change something. That's a wild feeling which very few of our people have in American businesses. They can actually impact what they do. They can actually change their own work space, improve their own life.

Now, when you announce and establish a structure like this, is every single one of your employees going to be wildly enthusiastic the first day? Hell, no. But enough will be to get you started. Everybody else is formally enrolled in the process, and that means that they can become part of it as soon as they believe you.

I'll give you one rule of thumb to judge your own process and the processes used by others. The question to ask yourself and other people is: If I work for your organization and I believe the speech that you gave last week, and I've thought of something that will improve what I do or what my section does, and I brought this idea to work with me Monday morning carefully nurtured to life, where do I go with it? Is there a place I can turn to with my idea and have it heard quickly and fairly?

Because if there isn't, you've got more work to do. If I get stepped on, if I get told to stuff it, that's my last idea for ten years. You've just lost me. And you can't afford to lose me or anybody else.

One hundred percent involvement means, when I come to work I've got some place to go because I'm enrolled in a quality team, and I can turn to my quality team and say, hey, when is our next meeting because I've got some keen ideas. I've got someplace I can go with that. I feel like I'm part of it. I can impact.

Emphasize both formal and informal communications, both internal and external. You've got to talk to your folks.

When we began the process at Paul Revere, we recognized that we were talking about a patriarchal insurance company—everything good and bad that patriarchal brings with it.

One of the bad things was that folks weren't talking across the layers and the boundaries. We wanted to break that up.

We established a program for ensuring that everybody is thanked. It's a very simple process. The first of each month, each of the top executives—24 vice presidents and above—received a "PEET sheet," a piece of paper bearing the names of two quality team leaders.

Their job, this month, was to go talk to those two team leaders. That's all, but we gave it a format. At first, did they all do it? No. Several months deep, the company president stopped me in the hall and said, "Are these guys doing those PEET visits?" because he was doing his.

I responded, "Not all of them." He said, "We'll take care of that." He had a monthly meeting of his expanded staff. "Here's a new rule, guys. When you've finished your PEET visit, give the sheet back to Pat Townsend filled out with your comments. Once a month, Townsend will give me a PEET sheet report. And we'll discuss it in here, if we have to."

At those expanded staff meetings, the first thing on the agenda every month was quality. The last thing was the PEET-sheet report. God help you if you hadn't found time to go talk to two people all month.

Along with communications, you also emphasize training. You can't just announce this stuff. You're going to have to train folks in it. You have to build some training courses. Tom Peters, who writes books on quality, has a nifty rule of thumb. He says that if your company is doing well, double your training budget. If it's not doing well, quadruple it.

It's an investment, not a cost. But you have to teach people what you mean by the word "quality," what you mean by the word "customer." You have to teach quality-team leaders what the process is, what the structure is.

Teach them how to run meetings in a participative and efficient manner. How to solve problems. You have to teach leadership, teach communications. These are not things that folks necessarily know intuitively. Give some format to them, give some continuity.

You'll have to do some measurement. Take the measurements that count, however. Don't just measure everything in sight, or you'll get a neat collection of charts that nobody will look at. Pick your spots. Measure important things. Do not use measurements as a weapon. For too many organizations, particularly manufacturers, measurements are used either as a weapon or as a religion. It's inappropriate as either.

Measurements are useful as a source of ideas and to check on progress. You can measure things and then turn the information over to your people and say, hey, 22 percent of this was unsatisfactory, do something about it. Then get the hell out of their way. Give them the authority to do something about it.

The last of these principles I want to give you today is to say "Thank you." For God's sake, say "thank you." When you build your program for recognition, gratitude, and celebration, keep one thing in mind. Different folks hear it different ways. Not everybody hears "thank you" the same way. We are different people. We are at different places in our own personal development. We hear it different ways.

Many of the programs you come across have really neat things they give to people. But they give everyone the exact same thing. They all get a plaque, or a this or a that. That's wonderful, and very fair. But it's terribly inefficient, because not everyone likes plaques.

I'll give you an example of what I mean by this. At one company I worked with, we said, "When you complete 10 ideas or have saved $10,000 for the company on an annual basis, you are a bronze team; 25 ideas, $25,000, you're a silver team; 50 ideas, $50,000, you're a gold team." I figured that $50,000 would last a year because 50 ideas, $50,000, that's a lot of ideas, a lot of money.

We started in January, and in April the first four teams made gold. And they said, "What's next?" So we added double gold, 50 more ideas or 50,000 more dollars; then we established triple gold and quadruple gold, and quintuple gold, and sextuple gold, et cetera. Folks do marvelous things if you give them the chance.

For thank-you's, always use your heavy hitters. The first two years, we only used the president of the company and two co-chairmen of the quality committee. Finally, with the third year, we loosened up and used a couple of senior vice presidents. But it's always somebody from the very top coming in to say thank you.

Each person will receive a lapel pin or charm. Now, for some folks that's really neat. It proves I'm good, and they said "thank you" to me. For other folks, it goes in the drawer with all of the ones from United Way. Some folks don't wear those things. Okay. Each person gets one, though, because I don't know who wants them. Each person gets one.

Each person also gets $20 in Paul Revere gift certificates good at about 50 stores and restaurants in town. We created our own currency. That gave folks the opportunity to spend it on whatever they wanted to. Why give them that and not cash? If you give them cash, it goes in their pocket and it's anonymous in about 15 seconds. But you give them a gift certificate, and they've got to spend that particular money consciously, they'll remember what they bought with it. They'll always remember that they got this thing because the company said "thank you." It's money with a memory, if you will.

Everybody gets written up in one of the company publications, gets his name in print. That's powerful, because I can highlight my name and send it on to my mother. "Dear Mom, the rumors you heard aren't true. I'm doing good work here." For some folks, that's what did it. For some folks, the fact that the senior vice president looked them square in the eye and said "thank you," that's what did it.

If they hear you say thanks, they'll do some more of it. This is a business decision we're talking about, a business process. You want them to have more ideas and improve more things. How do you do that? You say "thank you."

Now, you notice I have not used the word "incentive," because it's not incentives I'm talking about. It's a thank you for work performed.

Another idea we used is a Quality Coin. "The Big Q." On the back side it gives recognition for a job well done, and it's good for one meal at the Paul Revere cafeteria. That's all it is. Just a very quick thank you. Top managers are required to give out five a month. So you keep these on your desk, in a little stack. At the end of the month, the stack's got to be gone. What happens is, the vice presidents pick up networks, if you will, of informants. You know, you go in and tell him that Mary Jo stayed extra hours to finish that report. So the vice president goes out to Mary Jo and says, "Thank you."

Does it work? They've been giving these out for five years there now. Only 20 percent come back to the cafeteria. People saved them, because somebody actually said "thank you" to them personally.

Think about it for a second. When is the last time you consciously thanked an individual for something they personally did to help their organization?

I'd like to offer a couple of other thoughts.

One, please keep in mind that when you begin talking about quality in an organization, do not expect it to simply become part of the corporate culture immediately. That's just not going to be.

When you begin the process, people's questions tend to be, "What can you do for me?" Oh, good, we're having a quality process and George is going to quit screwing up and that will make my life easier.

The next step tends to be, "What can I do for me?" as folks come to realize that they're being given the ability to impact their own worklife and improve their own contribution.

The third step is, "What can I do for you?" Now, we start dealing with customers.

There is a fourth step, by the way. The fourth is, "What can we do together?" Then you really start working in partnership with your customers.

But keep in mind, you have to go through the first two steps – what can you do for me, what can I do for me – and only then will I worry about what I can do for you. Keep in mind also that this quality thing must be given attention separately. You're going to have specific meetings about quality to discuss it, to work on it, to decide how to do things. Most of the meetings that you hold in your organization are concerned with yesterday or today. Quality is talking about tomorrow.

If you schedule a meeting where the first 15 minutes are on normal staff business, and the last 15 minutes on quality, you know what's going to happen. There's going to be an emergency come up and you're going to do 29 minutes on today and one minute on quality.

Quality is an important enough topic to call for a separate time. So the question is, in a deadline-driven organization, "Do we have the time to do this stuff?"

You'll find the needed time once you make a commitment. Once you agree that quality is important, the time will be there.

We are talking about survival. You need to use all the brainpower on your payroll. The folks who work with you and for you are wonderfully creative people, marvelous folks. But too often, they're only marvelous outside the office. They'll be marvelous inside the office, too, if you give them the structure, give them the support, give them the reason to believe that you're serious about this.

Now, you've got about 10 minutes for questions and discussion. I'll answer any questions about this whole idea of quality.

Mr Guittar: In terms of what you're saying, is this what Deming would be telling us if he were here?

Mr Townsend: Deming's conversation would be more based on numbers. And he'd be far tougher on top management than I am. But the general theme, yes; he is very pointed in his comments about top management and their failure to understand and to put the program into action.

Mr Blodger: What do you think about an active suggestion program? You didn't mention that as one of the tools you use.

Mr Townsend: Suggestion programs have been around for years, and there is an organization called the National Association of Suggestion Systems, NASS.

At Paul Revere, we added up all the suggestions we had per employee and compared that with the national average for suggestion systems, and it's 350 to 1. The suggestions system is, to me, at the bottom of the list for these reasons:

It is hard for me to act on my own. I haven't got team support; I haven't any way to discuss this thing with others. I've got to write it out. The average response time for a suggestion is measured in months, weeks at best, so there's a tremendous time lag involved. I throw it up there for strangers to look at.

I think suggestion programs are a minimal approach. I want to go after people and say, "You, you're part of the quality process. We think you know things. Here's the process. Here is your encouragement. Here's your training. Please help us improve."

The thing is, when you start being conscious of quality and start looking at it, you're amazed at how much pops up.

Mr Winter: You talk about the required leap in faith. There is an expense involved in real development of quality. How do you sell stockholders on the need for that leap in faith?

Mr Townsend: There are a couple of thoughts there. To begin, it doesn't need to cost nearly as much as most consultants will charge you. There are some organizations out there that charge amazing fees, and I'd have a hard time justifying them to stockholders or anybody.

With too many consultant firms, basically, you buy them to come in and do it for you or to you. I have trouble with that. It's not going to work because the minute they leave, everybody says, "Well, good, the consultants are gone, now we can get back to work." No, you can't let it happen that way.

The investment really is not that big. It's real cost, but I think it's one that would be easy to justify. It won't be an outrageous sum unless you go with someone who charges an outrageous sum. If you try to do it passively by hiring others to do it for you, you've given up the game going in and you've spent an amazing amount of money.

The investment will always be returned. I mean if you screw up a quality process, you will at least get your money back probably five times. If you do it right, you'll get maybe 20 times your investment back. And there's enough documented proof that you can be very comfortable presenting that to stockholders.

Mr Guittar: It occurs to me that one of the things you have to do in a company is be awfully sure that employees are going to react positively. If you give a Quality Coin to somebody, his reaction might be, "Huh, big deal," and make fun of it. Then you're going to be reluctant to do that again.

Mr Townsend: If the coin was the only thing you did, you certainly run that risk. But as a part of the whole, where everybody knows this quality process involves everyone in the company and everybody knows that this executive is himself or herself on a quality team, that changes the picture. If I know that everyone else is involved in this thing, too, then it becomes a very logical piece of a whole.

Mr Guittar: I am reminded of American Airlines. I don't know how many of you get these little things from American, to be given to American's employees, a very special thank you to an employee. I've never given those to anybody, I just accumulate them because I don't know what the reaction would be. But American may be on to a great idea.

Mr Townsend: What happens to the employee when he gets them? What I want to know is, okay, I'm an American Airlines employee, somebody just gave me one. Is that it? If I get five, do I trade them in for a set of glassware? What's going on?

Mr Blodger: Maybe they know, I've never asked them.

Mr Townsend: Yes, but if American wants me to give them, they should tell me, too. What am I contributing to? What am I asked to be a part of?

Mr Blodger: Do you think an investment in membership in the...

Mr Townsend: What I know of the APQC, they're a good outfit with a good set of resources. They'd be worth talking to. They publish a series of things, and they deal with services as well as manufacturing.

Mr Finneman: I think with that I'd like to close this session. I'd like to thank you very much, Pat, for agreeing to be with us. I think you got us off to a quality start.

8

THE BALDRIGE

The Malcolm Baldrige United States National Quality Award has, in little more than a decade, established itself as the premier standard in quality, in large part because it is descriptive of desired results rather than prescriptive of required procedures. The criteria itself have been the subject of continuous improvement. The first set of criteria had a distinctly manufacturing bias which has since disappeared.

One of the authors of this book was on the original committee that drafted the Baldrige in late 1987 and subsequently served as an Examiner for two years. The other author was the primary author of Paul Revere Insurance Group's Baldrige application in 1988, the award's inaugural year. Paul Revere was one of two finalists in the Service Category (the other two categories being Manufacturing and Small Business) that year. No service company won a Baldrige until 1990 when FedEx was declared a winner.

Besides becoming an integral part of the American business landscape, giving people in organizations of all types and sizes a common vocabulary with which to transfer lessons across boundaries, the Baldrige has been the model for the national quality awards of over 40 countries.

Nothing about the Baldrige is magic nor is it difficult to understand (two of its strong points) so the hiring of a "Baldrige Consultant" should be done with care and common sense. "Pat Townsend on Choosing a Baldrige Quality-Assessment Consultant" offers guidelines to help in picking an adviser and reminds executives not to hand "responsibility to an outsider."

The other article in this chapter—"The Importance of the Baldrige to US Economy"—starts with a brief history of the Baldrige and goes on to describe the impact it had on the American economy between 1988 and mid-1996. Any nation can realize the same positive fiscal impact, although, as pointed out earlier, democracies have an automatic advantage.

PAT TOWNSEND ON CHOOSING A BALDRIGE QUALITY—ASSESSMENT CONSULTANT

The Malcolm Baldrige National Quality Award has become a coveted honor for organizations across the US—and a big business for consultants hoping to cash in on the latest quality craze. But beware of the bandwagon: Scratch the surface of this year's Baldrige consultant, and you'll find last decade's "excellence" consultant.

Many senior executives, having decided to go for the Baldrige gold, are now sorting through dozens of proposals offering exhaustive–and expensive–quality-assessment programs. So now what?

Do your homework, assess yourself By all means use the Baldrige criteria as a full quality audit for your company—*but do not hire a consultant to do it for you.* After all, the Baldrige application is written in plain English. And the director of the Baldrige Award program, Curt Reimann, often speaks at public seminars. Call the Baldrige office at 301/975–2036 to see when he'll be giving his next presentation. Contact the sponsoring organizations and reserve seats for your top managers. Before going, the group should invest at least one full day reviewing the Baldrige Award application and deciding what questions to ask. On returning, they should prepare to follow through intensively and in person.

Only if top managers *conduct their own assessment* can they truly understand the strengths and weaknesses of their organizations. And only if they are seen to do it themselves is anyone else in the company going to believe that they are serious about quality.

Get advice, but do the work yourself Think of quality as a do-it-yourself project. Consultants can give advice.

They can save time drafting your plans. They cannot, however, do the job for you. There's no substitute for your own judgment and involvement.

With that in mind, here are five guidelines for selecting a quality consultant—whether your goal is to win a Baldrige or "simply" to launch an internal quality initiative. The winning candidate should:

- Explicitly urge 100 percent involvement—formal, literal participation—of everyone on the payroll. The proposal should include a way for this to happen. The point is not just that every employee is affected by the quality process. Rather, every employee must *have a way to affect the process.* Be sensitive to the difference.
- Let your organization reproduce, revise and update instructional material so your training department can truly take control of these courses.
- Indicate a clear time frame for completion of the engagement. If he or she is vague on this point, ask the age of the consultant's youngest child; then calculate the number of years it will take the kid to complete college. That's how long you can expect the process to last.
- Make clear that your managers remain in charge of the process. Dump any proposal that does not *specifically refuse* to do top management's job for them.
- Address the fact that quality improvement is an emotional as well as intellectual endeavor.

Any proposal that can't pass these guidelines is probably not up to par. And one more caveat: If the consulting firm insists that every person in your organization be trained before the process begins, or that its own "facilitators" be on hand for every Quality Action Team meeting (or whatever), don't just toss the proposal—burn it.

Remember, the Baldrige is not a fool-proof lesson plan for success. It is descriptive, not prescriptive: It provides the questions that need to be answered, not the answers themselves. As such, it is a marvelous assessment tool. Don't blunt its power by handing responsibility to an outsider.

On Achieving Excellence, *Aug. 1991.*

THE IMPORTANCE OF THE BALDRIGE TO US ECONOMY

To fully appreciate the impact of the Malcolm Baldrige National Quality Award keep in mind two things: the mood among American businesses in the early-to-mid 1980s and the number of "Baldrige clones" now in existence.

Memory of the business mood in the 1980s underlines the importance of the Baldrige to the American economy; recognition of its clones testifies to the versatility and strength of the award worldwide. If there was ever any doubt as to which quality award is the *first among equals*, Japan ended the discussion by rolling out a Baldrige–based Japanese Quality Award in 1996.

The wake up call Students of American business history can relish the irony. Less than two decades earlier, the 1980 NBC television white paper, *If Japan Can Do It, Why Can't We?*, sounded an alarm over the magnitude and nature of the Japanese challenge to the American economy. By the mid-1980s, American business people were being told by customers, analysts, and journalists that they were in serious trouble—clueless as to how to stem the flow of Japanese automobiles and electronics into American markets. Pilgrimages to Japan to learn how they did it were a regular feature on the calendars of American business executives.

Concurrently, the private sector made several attempts to establish a national quality award along the lines of the Deming Prize, first awarded in Japan in 1950. Seen as a key element in Japan's meteoric rise from being the source of *all-things-cheap-and-breakable* to being the acknowledged leader in quality, the Deming Prize had rallied the Japanese business community, providing a common vocabulary and an agreed-upon model.

Moreover, Deming Prizes were announced on national television, reportedly drawing large viewing audiences and making the award a major commercial plus.

Other Americans looked closer to home for inspiration. Tom Peters and Bob Waterman wrote *In Search of Excellence*, the first serious (and popular) effort to tell the American business community that all was not lost and that there were American exemplars of excellence that could serve as models for rejuvenating the economy.

And then came the Baldrige

The basic legislation for the Baldrige was introduced in Congress in early 1986, quickly relegated to a committee and, equally quickly, pigeonholed. The proposed legislation wasn't, however, completely abandoned. By spring of '87, occasional discussions on how to implement the idea resulted in the decision to have the National Bureau of Standards design and administer a national quality award—if the legislation ever became law.

Sadly, the death of Secretary of Commerce Malcolm Baldrige proved to be the catalyst for making the award a reality. Baldrige was both liked and respected by members of both political parties, besides being a highly effective member of the Cabinet and a personal friend of President Reagan. In his 60s, he was—to use an old phrase no longer in favor—a *man's man* who still occasionally competed as a rodeo cowboy.

On July 25, 1987, he was mortally injured in a riding accident (he had been inducted as a *Great Westerner* by the Cowboy Hall of Fame in 1984). Soon after his death, the half-forgotten legislation, resurrected with Baldrige's name attached, swept through Congress on a voice vote.

The role of NIST and Curt Reimann Suddenly, the nation had the Malcolm Baldrige National Quality Award with the National Bureau of Standards as its administrator. NBS (now called NIST, the National Institute for

Standards and Technology), a branch of the Commerce Department, was an inspired choice: Its employees are, undeniably, the least political group of civilians on the government payroll. In addition, the organization had in recent years begun to wrestle with the fact that the use of measurement in industry appeared to be moving from finished product inspection to in-process prevention and correction.

And, most fortuitously, Dr. Curt Reimann was at NBS. He had become a member of the American Society for Quality Control a couple of years previously and had been instrumental in getting NBS enrolled as an institutional member of the ASQC. As a result, he had some knowledge of the players in the burgeoning field of quality, and he was aware of their range of preferences with regards to criteria for a national quality award. Happily, the fact that Reimann had no known allegiance to any one guru or methodology allowed him to be an *honest broker* in the discussions that followed.

When the legislation passed, NBS and Reimann could be sure of three things:

1. President Reagan could most likely be counted on to present the award named after his good friend.
2. Reagan was leaving office at the end of 1988.
3. There was little or no agreement about how to proceed.

Speed bumps along the implementation path... To complicate matters, there were at least two major types of rifts in the quality community: one between the *quality people* and the *productivity people*; and a series of divisions between the disciples of the various recognized gurus.

To complicate matters even further, the president of the ASQC had blasted the idea of a national quality award in an editorial in the ASQC's journal, *Quality Progress*, in March 1987.

There was even a minor battle between the *quality control* people and those who preferred the term *quality assurance.* And the incredibly short time frame meant that eveyone was going to have to cooperate—with Reimann and with each other.

Calling on Dr. Deming... In the end, it was impossible to satisfy everyone—especially Dr. W. Edwards Deming. During the process of inviting everyone in under the *big tent* he was constructing, Reimann personally called virtually everyone who had been active in some attempt to build a national quality award and/or was a known voice in the field. One of the first people he called was the redoubtable Dr. Deming.

Within a minute of telling him who he was and what was going forward as a result of the new law, Reimann was informed in no uncertain terms that:

- There was no organization in America good enough to receive such recognition...
- And even if there were, there was no one in America knowledgeable enough to make such a judgment.

Deming never changed his mind and he never accepted the award's views on goals and bench-marking and recognition—to name just a few.

Juran, Feigenbaum and Crosby... The two other major gurus (Doctors Joseph Juran and Armand Feigenbaum) supported the award from the outset, offering constructive ideas and their personal endorsements.

Phil Crosby withheld his endorsement, objecting to—among other things—the idea of self-nomination/application. He proposed that customers nominate recipients, overlooking the fact that customers are not in a position to evaluate the processes that create customer satisfaction. Self-assessment and feedback are major benefits of the award; neither result from a customer nomination. While Crosby continues to snipe at the Baldrige, several Baldrige winners—including the two 1995 winners —point to him as part of their "quality roots."

Getting the Baldrige on the street By the time that he convened his first official meeting on September 24, 1987, Reimann already had a basic outline of the award application, to include the seven categories that have become one of the distinguishing characteristics of all Baldrige clones.

The criteria had to be *on the street* by mid-January if a full cycle was to be possible before Reagan left office. To make the rush for the first award cycle more palatable, Reimann promised that the award's criteria and procedures would themselves serve as a model of continual improvement through an annual review and modification procedure. If a particular faction was unhappy over a specific detail, they knew they would get a crack at revising it the following year.

Reimann's personal schedule was keyed to a meeting in late October with the new Secretary of Commerce William Verity, who proved to be a true advocate of the award, offering invaluable support. Prior to October, Reimann, a senior member of the NBS staff, had never made a presentation to the Secretary of Commerce, much less on a topic that had presidential interest. He has since said, "Terror helps to set priorities." The only reason that he and his small staff only worked seven days a week was that they couldn't manufacture an eighth day.

The NBS and Reimann succeeded The Baldrige Award Criteria was available in January, 1988; 12,000 applications were requested that first year; and President Reagan made the presentations that fall.

Reimann's role cannot be overestimated... As mentioned above, Reimann defined the seven categories that have proven so durable. He credits his background as a chemist and scientist for having trained him to look for fundamental truths that would not ebb and flow with changing opinions and personalities. When asked, "Why seven?" Reimann paraphrases an old Volkswagen commercial explaining why the VW had four forward gears by saying,

"Because six was too few and eight seemed like too many."

And Reimann built lasting bridges between people... The fact that there is a coherent quality community in the United States and a good share of the world, a community with a fairly consistent view of what is meant by the word *quality* and how to achieve it, is directly attributable to Reimann's Baldrige efforts.

From negotiating the original definition of the Baldrige to winning near-unanimous support to guiding the growth of the award through its first several years, he was of extraordinary importance.

Reimann has since retired, although he is still hanging around NIST, answering questions and offering opinions. (Dr. Harry Hertz, a friend and colleague, is the current Director of the Baldrige program.)

Growth and strength of the greater Baldrige One of the signs of growth and strength of the Baldrige is the number of Baldrige clones throughout the United States in states, counties, cities, businesses, and non-profit organizations such as the United Way of America—not to mention foreign countries.

While skeptics question why the number of applicants for the Baldrige remains flat over the years, Reimann points out that the number of applicants for Baldrige-based awards in the United States continues to grow.

Self-assessments grow... Last year between 700 and 800 organizations went through a process of self-assessment by filling out an application and received feedback from a minimum of three outside examiners/consultants. The implications are enormous. Even if an organization applies only once and drops out after the first examiner assessment, they have a much better idea of what they don't know and/or don't do.

Combine the growing number of Baldrige and Baldrige clone applicants with the number of organizations that have signed up for classes by Baldrige winners and you have the *American Quality Revolution*.

Other measures The Baldrige is not, of course, the only formal, outside–measured, public recognition for quality. The Deming Prize and ISO 9000 are also universally acknowledged. While all intend to focus on customer satisfaction and continual improvement, the three can be compared briefly:

- The Deming Prize lays down a set of prescriptive criteria, a *road map* of the only route to quality.
- The Baldrige criteria relies on a descriptive criteria, a combination destination and compass with a request for information about which routes have been chosen.
- ISO 9000 requires evidence that a trip, almost any trip, is made without requiring assurance that the destination has been reached.

The contrast between the three explains why Baldrige winners are invaluable to the rest of the business community: The Baldrige encourages a wide variety of approaches to reach the common goal of customer satisfaction, as long as the emphasis is on processes that insure continual improvement. After eight years of winners, the selection of *lessons-to-be learned* is a virtual smorgasbord of ways to make money—assuming the listener is willing to put in the necessary work.

The financial impact of quality is beyond reasonable challenge After Dr. Juran made a casual comment that he was sure that buying stock in Baldrige winners was a solid investment, several journalists decided to check it out. Using 20–20 hindsight calculations, no matter how the answer was computed (just whole company winners such as Motorola and Federal Express; companies whose subsidiaries won, such as AT&T or GM; everyone who received site visits), the results were consistent: Earnings were significantly ahead of the Standard & Poors average for the stock market as a whole.

A cynic would observe that for an American business executive not to know that pursuing quality is financially sound requires a willful misinterpretation of widely available data. (Unless you confine your reading to the *Wall Street Journal* which never has figured out that quality makes money.)

The Baldrige quickly established itself as a major force in American business Early winners such as Motorola and FedEx became passionate corporate missionaries for the cause, spreading the word to anyone who would listen. While the original law specified that winners would be required to teach, no one dared to dream that the winners would take to that task so enthusiastically.

The role of the examiners... Reimann, his NIST team, and the various members of the quality community also made a pivotal decision when they opted for a rotating group of award examiners rather than a stable of in-house experts. In practice this has meant that trained examiners go back to their own firms and continue educating the American workforce. Since the primary objective of the Baldrige is to spread information about the concepts of quality, this ripple effect is exactly what the legislators had in mind.

Each year, there are approximately 2,500 volunteers trained as examiners for the Baldrige and all the Baldrige clones in the United States, creating a pool of quality experts. Other nations that have adopted Baldrige-based national quality awards have reported a similar ripple effect.

Deming may well have been right in thinking that the average JUSE (Japanese Union of Scientists and Engineers) *consultant* who works full-time as a professional consultant is more knowledgeable than the average Baldrige examiner who holds a part-time position for two years. JUSE consultants, however, typically determine winners of the Deming Prize from among the three-to-five Japanese companies who apply each year.

Table 8.1 Eight years of Baldrige winners: 1988–1995...

Ames Rubber Corporation (1993)	GTE Directories Corporation (1994)
Armstrong World Industries Building Production Operations (1995)	IBM Rochester (1990)
AT & T Consumer Communications Services (1994)	Marlow Industries (1991)
AT & T Network Systems Group (1992)	Milliken & Company (1989)
AT & T Universal Card Services (1992)	Motorola Inc. (1988)
Cadillac Motor Car Company (1990)	The Ritz-Carlton Hotel Company (1992)
Corning Telecommunications Products Division (1995)	Solectron Corporation (1991)
Eastman Chemical Company (1993)	Texas Instruments Incorporated–Defense Systems & Electronics Group (1992)
Federal Express Corporation (1990)	Wainwright Industries, Inc. (1994)
Globe Metallurgical Inc. (1988)	Wallace Co., Inc. (1990)
Granite Rock Company (1992)	Westinghouse Electric Corporation Commercial Nuclear Fuel Division (1988)
	Xerox Corporation–Business Products & Systems (1989)
	Zytec Corporation (1991)

Award winners in alphabetical order.

Given their year–round status as outside consultants and their relatively small numbers, they are incapable of having a long range impact equal to that of the Baldrige examiners.

THE 1995 BALDRIGE WINNERS

Corning Telecommunications Products Division and Armstrong Building Products Operations—the two 1995 Baldrige Award winners—have already stepped up to the plate and assumed their roles as teachers of quality. Even though they could fulfill their statutory requirements by taking part in a *Quality Quest* gathering and four regional conferences, they have chosen to follow in the footsteps of their Baldrige predecessors and make their people and their processes available to as many organizations as is reasonably possible.

Bob Knezovich, Armstrong BPO's manager of quality management, said in May that there were already over 500 requests for speakers, visits, information, and interviews registered in their system, including calls from around the world.

In the first 3½ months after the awards ceremony, Catherine Shaw, Corning TPD's communications coordinator for the Baldrige, reported that monthly *Winning Through Excellence Quality Seminars* have already proven to be very popular. During the same period, TPD personnel gave 16 speeches at locations ranging from Oregon to New York, and there were 32 more speeches scheduled for the months ahead, including presentations in France, Finland, Chile, and Puerto Rico. With more invitations arriving almost daily, Shaw said, "It has been a wild ride," since winning the Baldrige. She also opined, "It is exciting to share your story with people who want to hear it."

The 1995 winners have stories worth sharing... In interviews, both Corning TPD and Armstrong BPO executives were eager to explain what makes their efforts unique.

Corning TPD Corning TPD takes pride in the fact that they truly have done what many organizations use as an excuse for not pursuing a formal quality effort: They've made quality a part of everything they do. When TPD says, "All of our people do quality work all the time," they can prove it.

In fact, the word *quality* shows up relatively infrequently in TPD literature and conversations. There are, for instance, no *quality classes* on their training curriculum. For that matter, there is no *quality manager* on the

TPD wiring diagram. There are, however, plenty of business classes—enough so that everyone on the TPD payroll is urged to spend 5 percent of their work schedule each year in class.

In short, the decision was to focus on TOTAL BUSINESS rather than TOTAL QUALITY, while using all manner of quality tools.

Involvement of Corning TPD employees... Continuing a trend that became evident with the 1993 winners, TPD's quality efforts involve every person on the payroll, even without a formal structure (e.g., a network of teams or a suggestion sytem). Everyone is trained for and expected to take part in improving quality.

Because quality really is the way they do business, and because the Baldrige philosophy and criteria are imbedded in the every day practices of TPD, no one is hesitant to put forward ideas or to suggest modifications and offer help in implementing someone else's idea. There are a number of cross–functional teams, but their individual progress is not tracked at any central point.

Rewarding and recognizing participation... Corning TPD has a very active program of rewards and recognition to encourage active participation. According to company vocabulary, *rewards* refer to financial bonuses and *recognition* refers to more personal and more frequent acknowledgement of achievements. The rewards and recognition efforts are part of the overall TPD drive to insure that all employees feel valued. Only then, the company management believes, can they do their most effective work.

Key Result Indicators... One aspect of the financial reward system is a variable compensation bonus program which reaches all 1500 employees and, depending on company–wide progress against six *Key Results Indicators*, can result in a bonus of up to 10 percent of every employee's pay. Each year, everyone gets the same percentage bonus, be it 0 or 10 percent. The *Key Results Indicators* are re-set each year with the previous year's results as the benchmark. In other words, if Corning TPD does just exactly as well one year as it did the previous year, there is no bonus. Getting better is what counts, not holding on to the status quo.

There are also a number of instant recognition devices available which make it possible for employees to thank each other for exemplary performance, and three times a year, there is a recognition breakfast. This latter is run by the non-management employees and is designed to spotlight individuals and teams who have taken a turn in leading the way toward improving business processes.

Everyone on the dance floor... TPD Vice President Bob Forrest's imagery for participation is well known throughout the organization. He compares the early days of a quality effort to a high school dance:

- Some people can't wait to strut their stuff, getting out on the floor almost before the music begins.
- Others get up and dance after they see other people enjoying themselves.
- Still others stand in the hallway debating whether or not to go in.
- Then there are people out in the parking lot who won't admit, least of all to themselves, that the reason they are there is that they wish they could dance. So, instead, they cause trouble.

Forrest wants everyone on the dance floor, even if that means evangelizing and teaching and prodding the employees who are slow to *get it.* The dance isn't truly a success unless everyone participates.

Jerry McQuaid, Corning TPD division vice president, a frequent spokesman for their quality effort, affirms that quality is not something you do in addition to work, but rather how you work. He admits to having a reaction that falls someplace between amusement and horror at some of the questions he receives on the rubber chicken circuit as his audiences grapple with the question, "How will we ever find enough time for quality?" He says that some people simply will not accept that quality is not something "you bolt into the side of the business."

Leadership as a central focus... Leadership is a continuing focal point at TPD. McQuaid's and Corning TPD's stated philosophy is that "Leadership is a behavior, not a position." Leaders at TPD are responsible for establishing linkages between visions and what people actually do by devising systems that integrate goals and efforts.

The year 1995 was not the first year that Corning TPD applied for the Baldrige... In 1989, they applied and received a site visit, but McQuaid admits that they really weren't then fully committed to the ideas in the Baldrige application. They were a good company; they were making money; they were obviously doing a number of things well or they never would have gotten the Baldrige Award site visit.

By 1993, however, senior management had come to realize fully that the Baldrige offered a coherent framework for all the smart things they were already doing, albeit in a disjointed manner. The criteria also helped pinpoint where things needed to be done, according to McQuaid.

By 1995, Corning TPD was ready to have the Baldrige examiners take another look and the company applied again. No one person had the sole job of putting the application together. As has been true for virtually all Baldrige winners, completing the applications was not particularly difficult once the company is ready—it was just a matter of writing it all down in the right sequence.

Armstrong Worldwide Building Products Operations

Co-winner in 1995, Armstrong Worldwide Building Products Operations came to quality via a different route. Henry Bradshaw, the president, is credited with being the catalyst behind Armstrong's wide-ranging vision statement. He tells a story about how he asked directions to the men's room after a long afternoon of negotiations and tea-drinking at a cafe outside of Quangzhou, China. He followed the directions down a long, rickety walkway to an outhouse. While in the outhouse, he noticed a piece of Armstrong ceiling tile nailed overhead. From that experience came Armstrong's vision:

Our products will heighten the beauty, comfort, and safety of buildings and homes everywhere.

And he does mean everywhere.

The beginning of Armstrong BPO's use of the Baldrige criteria as an assessment tool is sometimes traced to another story told by William B. "*Bo*" McBee, who was the corporate director of quality management for Armstrong Worldwide Industries in 1989. When a member of the board of directors asked him how the quality efforts were going, he simply answered, "Fine, thanks." After returning to his office, McBee began to think about his response and decided that it fell a bit short of being complete and informative. He decided to take a detailed look at options that might help him organize a more convincing response.

Crosby roots... At the outset of their quality efforts some years before, Armstrong, in common with Corning TPD, had found Phil Crosby's teachings most helpful.

Note: No matter what level a Baldrige applicant reaches, there is always a feedback report. The higher the level reached, the more detailed the feedback simply because a greater number of examiners have looked at the application and, in the case of those who receive site visits, the organization itself. The Baldrige feedback report is arguably the best bargain in consulting in America. Armstrong BPO's 1993 Baldrige feedback report played a major role in the maturation of their quality effort over the ensuing two years. When in 1995 they applied again for the award, they won.

In fact, Armstrong continues to use a modified version of Crosby's 14 points in their process. By 1989, it was time for something more detailed, more attuned to improving all aspects of the business.

Quality focus as a business decision... To their great credit (and they share this distinction with their fellow-award winners at Corning TDP), Armstrong BPO was not forced to *do quality* because of a crisis in their business. They simply made the rational decision to proceed because it made a great deal of sense to them as a business decision capable of both strengthening the bottom line and making both customers and employees happier. As Bob Knezovich says, the Armstrong leaders realized that quality was " the way into the future"; although like Corning, Armstrong tends to use phrases such as *business excellence* rather than *quality* to describe what they do.

They're still working at improving involvement... But it wasn't automatic. Knezovich admits that many executives were slow to accept the idea of getting into something that, like a personal physical fitness program, promised great things in the long term for consistent concerted effort from this day forward.

Knezovich also points out that Armstrong BPO, despite being a Baldrige winner, is still working on involving every single person in the organization in continually improving what they do. He puts the current percentage of active involvement "in the 90s" and says that 100 percent are touched by the process, even if not currently signed up as an official participant in the effort.

An internal quality award... To help the various divisions of Armstrong come to grips with the idea of quality, an *Armstrong Quality Award* was established—a Baldrige clone. The self-assessment component was invaluable in helping divisions both identify strengths and uncover weaknesses. In 1992, Building Products Operations won the Armstrong Quality Award; that same year,

Armstrong's Floor Products Operations received a site visit from the Baldrige Examiners. In 1993, Building Products Operations applied for the Baldrige and received a site visit and a detailed feedback report.

Participation in the award ceremony... Ever since the second Baldrige Award ceremony (the first was conducted by President Reagan in the East Room of the White House and the award was not announced in advance), winning companies have been encouraged to bring a group of employees to the award ceremony.

Armstrong BPO not only brought along employees, it brought along a number of customers. In fact, a major thrust of their quality effort is to deeply involve both customers and partners, such as trucking firms, in grading and upgrading everything they do. Customers, for instance, sometimes act as judges for Armstrong BPO in-house team competitions.

Through business excellence rooted in quality tools, Armstrong BPO has kept its growth in costs below the rate of inflation, managed its force reductions through attrition, and made a great deal of money. And while, unlike the Deming television specials, there is relatively little fanfare for the Baldrige, company officials admit that they could not have afforded to buy the tremendous positive publicity that they have received from winning the award.

Focusing National Attention on the Baldrige

Presidential attention to the award has been instrumental in creating momentum for the American Quality Revolution. The Baldrige has been awarded during the administrations of three presidents: Ronald Reagan, George Bush, and Bill Clinton.

The fact that Reagan himself presented the inaugural awards helped to pique public interest. Reagan went 1 for 1; Bush went 3 for 4 (with the one time he missed being the year that he was in Spain, kicking off the scheduled-at-the-last-minute Arab-Israeli peace talks). On that occasion, Vice President Quayle stood in for the President.

Disturbingly, President Clinton has been somewhat cavalier in his treatment of recipients.

- In 1993, President Clinton presented the award after changing the time of the presentation from afternoon to morning on the Friday before the Monday ceremonies, making life more than a bit challenging for those who were scheduling events before and after the ceremony.
- In 1994, Vice President Al Gore did the honors.
- There was no ceremony in 1995. In February 1996, Secretary of Commerce Ron Brown presented flags to the CEOs of the two companies who won 1995 Baldriges—the trophies were presented to the CEOs some weeks later by President Clinton at a small luncheon at the White House.

By conducting a private ceremony five months (and approximately 17 date changes) after the awards were announced, the Clinton administration insured that the 1995 Baldrige winners and the Baldrige Award itself received the least amount of media attention possible. If that private luncheon is counted as a *real* ceremony, President Clinton's record for participation in Baldrige recognition ceremonies is 2 for 3. Counting that luncheon, however, requires the 1,500 people who attended the earlier gathering—including the 120 or more employees from the winning companies—to wonder exactly what they attended.

Has the government cooled off on the Baldrige? The President's behavior is symptomatic of the current government indifference to the sucess of the award. The Baldrige, although widely accepted in the American business community and throughout the world has, for reasons difficult to discern, received surprisingly little support of late from either political party.

Pilot programs for developing and conducting Baldrige Awards for healthcare and for education were initiated in 1995, but had their funding withdrawn before the year was over. Despite the fact that the Baldrige was a major tool for rejuvenating American business, there are no healthcare or education Baldrige efforts in 1996—although no one disputes that the healthcare and education sectors could use improvement.

The Paradigm has Shifted

Notwithstanding the doubts voiced in the media and uneven support from the executive and legislative branches of our national government, the Baldrige will retain its position of importance, a position earned by being perhaps the major factor in positioning American business for the 21st century. The Baldrige didn't just shift the paradigm for American business—it defined a whole new way to go about doing things. As a result, business communities throughout the world once again can look to America to learn how to get things done.

Journal for Quality and Participation, *July–Aug. 1996.*

4

IF THEY CAN DO IT...

The Military as a Benchmark
The Paul Revere Process
Industry-Specific Examples

In large part because a quality process, when *done correctly*, involves every person in an organization and, subsequently, affects the lives of every person in that organization, there are almost as many aspects to it as there are to the human personality. Quality, like a tapestry, is woven of many threads. Any insight that illuminates how to make relationships between individuals—or between individuals and groups or between groups and groups—more productive contributes to the final design.

This section contains examples of how specific organizations have mobilized people for action. The first three articles focus on leadership in the United States military and explore the application of the principles to business. The next four detail the first five years of the Paul Revere "Quality Has Value" process. Closing the section are four articles that look at quality in specific settings: health care, communications and computers, software, and the United States Marine Corps.

9

THE MILITARY AS A BENCHMARK

The concept of a "benchmark" in the context of a quality process was introduced to the American business vocabulary by the Xerox company, winner of a Baldrige Award in 1989. Aware that their company did not handle telephone calls particularly well, Xerox (makers of duplicating machines) visited L.L. Bean (providers of casual clothes) to learn from L.L. Bean's much lauded telephone procedures. The company then adapted the "best practices" they observed to its own operation, a decision which was a key factor in their back-from- the-brink-of-disaster recovery in the late 1980s.

The military, especially the military of a democracy, can likewise serve as a source of information on leadership. This one skill, leadership, can be studied in isolation. With appropriate adaptations, without having a mission in any way similar to the military, best practices can be introduced into an organization looking for ways to improve.

The first article, "Leadership: An Ancient Source for a Modern World," was originally printed as one of a trilogy. The other two—"Participation: Starting with the Right Question" and "Measurement: Neither a Religion nor a Weapon"—can be found in Chapters 4 and 5 respectively. The eleven Leadership Principles included in this article are an excellent example of the transferability of techniques from the military to the business environment.

"The Three Priorities of Leadership" focuses on a consideration often overlooked in civilian organizations. Like so much of quality, the concept is simple: You must consciously set priorities. The execution, however, is difficult since all three principles are defined by the US Army. The leadership principles included in the other two articles in this chapter are those spelled out by the U.S. Marine Corps.

"What Military Can Teach Business About Leadership" introduces an insight that may catch some readers off-guard: Leadership is a subset of love. The eleven Leadership Principles introduced in the first article in this chapter reappear, this time matched with eleven Principles of Love.

LEADERSHIP: AN ANCIENT SOURCE FOR A MODERN WORLD

Virtually everything written on the topic of quality addresses at least one of three themes: leadership, participation, or measurement. Each is important in unique ways, but they have one thing in common. Together they contribute directly to the goal of all quality improvement: customer satisfaction. This article begins a series of three that will present what might be considered controversial looks at each area, beginning with leadership—where all quality must begin.

Among the traits that we admire in our leaders is the willingness to assume risk. Often left undefined, however, is the nature of the risk. We fail to differentiate between those times when a leader puts another's assets at risk—whether money or prestige or ego—and when a leader puts personal assets at risk. There is a difference; and in failing to differentiate, we short-change the concept of leadership. It isn't all that tough to put someone else's money or future in jeopardy, as legendary executives have proven repeatedly. Only when a top level manager is willing to put himself or herself at risk—their ego as well as their funds—is true leadership approached.

What criteria define the difference? It's more than identification with the success of a project. It's more than mastering a set of skills. It's an ephemeral combination of both caring and competence, beginning with knowing what to care about and when, how to direct and how to get out of the way. It's about people both in the singular and the collective. It's subtle and it's simple.

Perhaps a good place to start would be to look at the difference between management and leadership. That much, at least, is straightforward: managers care that a job gets done; leaders care that a job gets done, *and they care about the people who do the jobs.*

It's not a fluke that organizations are recognizing how important good leadership is to quality. Quality requires leadership—not management. Management is a subset of leadership in the same way that productivity is a subset of quality. Only the largest concepts produce the maximum results. Only when someone is ready and willing to incorporate courage and judgment, emotion and continual personal growth in their understanding of the task ahead can he or she move from management to leadership—or from productivity to quality.

Having defined leadership as more than mastering a set of skills, it is time to backtrack and say that mastering a set of skills isn't a bad place to start. Unfortunately, a growing library of books on the topic of leadership urge business men and women to move from manager to leader without any attempt at skill building. These leadership books are biographical in nature, either fullblown descriptions of the life and times of a single person or collections of biographical sketches of men and women commonly accepted as leaders. They are not the best way to learn to do the job yourself.

Starting a study of leadership with such personality-dependent books is like asking someone to hone his abilities to perform a task from looking at the results of another's efforts. Imagine, for instance, that you happen to wake up at 3:00 AM with the determination to become an expert (despite a total lack of knowledge) on the topic of physics and, indeed, to become a world-class physicist. Suppose further that, upon searching your random collection of books, two possibly relevant books are found: a high school text on physics and a biography of Albert Einstein. While the biography of Einstein might increase your desire to emulate the man, the textbook would provide you with the opportunity to begin to measure your aptitude for physics. Admiring Einstein is not enough.

That is why we are suggesting a text on basic leadership that may at first seem outrageous. At the very least, it will run counter to more than a few, possibly cherished, stereotypes. One profession, and one profession only, has formally studied the topic of leadership for at least 2500 years and has made a practice of transferring that knowledge to 18 year olds. The theory has been tested in a series of experiments where the penalty for error has been quick and severe. The results have been meticulously recorded and are a matter of public record. Even so, as a source of leadership techniques, this knowledge is virtually untapped.

The profession is the armed forces. What follows is a synopsis of several offerings on military leadership that appear in our book *Quality in Action*.

Conceding the stereotype immediately, when many people hear the phrase "military leadership," they quietly write it off as an oxymoron. "Military leadership" and "authoritarianism" are synonymous in many minds. When dictatorial practices in the military are trumpeted across headlines, it is well to remember that it is the *unusual* that gets reported—that's why it's newsworthy. Like so many stereotypes—the idea that all females are incapable of being high-level executives comes to mind—it is a stereotype built on the well-publicized failures of the few, rather than successes of the many.

The reality is somewhat different. At its best, military leadership is caring, participative, and lean. It relies on trust and purposeful delegation of authority to the lowest appropriate level. It is, in fact, a valuable resource for anyone intent on initiating or strengthening a quality process.

It is also easily accessible. Two of the best American examples are the United States Army's *Field Manual* 22–100 and the United States Marine Corps' *Guide book for Marines*. Both texts are public documents; both are written in clear language, what is sometimes called "standard prose." You will find counterparts of these manuals in virtually every country in the world.

What's in the manual? For one thing, refutation of the stereotype that the military accepts behavior in its leaders that civilians would never tolerate. FM 22–100 is quite clear on this point:

"Some people think that a leader is using the directing style when he yells, uses demeaning language, or threatens and intimidates subordinates. This is not the directing style. It is simply an abusive, unprofessional way to treat subordinates."

The text recognizes that there are legitimate uses of the authoritarian or directing style of leadership when practiced with courtesy and clarity. It is a valid and appropriate style when three conditions prevail: you are out of time, you have all the information you need, and *your people have high morale.* The reason your people have high morale, of course, is that you don't do this to them very often. An authoritarian leader retains both the authority and the responsibility for accomplishing the task.

Two other styles of leadership are explored. Participative leadership is demonstrated when the leader calls for the opinions of his or her subordinates and engages them in discussion about the components and alternatives. Please note that the leader still makes the final call and retains authority equal to responsiblity. By asking subordinates for their input, there is a positive impact on morale.

The toughest type of leadership to actually practice is delegative leadership. Here the leader delegates not only the task, but the authority to accomplish the task. Please note that it is authority—not responsibility—which is delegated. Responsibility always remains with the leader—no matter what style of leadership is chosen. That's what makes delegative leadership so unparalleled; the leader trusts subordinates enough to say, "I know you can do the job, but if something goes wrong, I'll back you." It's a tough thing to do, a risky thing to do—risky for the leader. But what a fabulous environment in which to learn. And delegative leadership has the advantage of leaving the leader time to deal with tasks that can't be delegated.

What is best about the manual is that it models what it wants most to encourage: flexibility and independent thinking. Beginning its discussion of the principles of leadership, the Marine *Guidebook* invites readers to draw their own conclusions.

Eleven leadership principles are set forth just for the sake of discussion. You may want to add or delete some. That's OK. We're not concerned as much about the words and phrases as we are about their application. They're all commonsense items, anyway. When you get right down to it, a discussion of leadership is only common sense with a vocabulary.

The eleven principles are easily translated to civilianese without loss of meaning, so we'll go through them one by one:

1. Take responsibility for your actions and the actions of your people The leader alone is responsible for all that the unit does or fails to do. That sounds like a big order, but take a look at the authority that is given you to handle that responsibility. You are expected to use that authority. Use it with judgment, tact, and initiative. Have the courage to be loyal to your organization, your people, and yourself. As long as you are being held responsible, be responsible for success, not failure. Be dependable.

2. Know yourself and seek self-improvement Evaluate yourself from time to time. Do you measure up? OK, you don't, admit it to yourself. Then get busy. On the other hand, don't sell yourself short. If you think you are the best at your level in your department, admit that to yourself. Then set out to be the best in your division. Learn how to speak effectively, how to instruct, how to use all the procedures and equipment that your unit might be expected to use.

3. Set the example If you are in a management position, your subordinates are already looking to you for a pattern and a standard to follow. No amount of instruction and no form of discipline can have the effect of your personal example. Make it a good one.

4. Develop your subordinates Tell your people what you want done and by when. Then leave it at that. If you have managers subordinate to you, leave the details up to them. In this way, kill two birds with one stone. You will have more time to devote to other jobs and you are training another leader. A leader with confidence will have confidence in subordinates. Supervise, and check on the results. But leave the details to the person on the spot. After all, there's more than one way to skin a cat. And it's the whole fur you're after, not the individual hairs.

5. Ensure that a job is understood, then supervise it, and carry it through to completion Make up your mind what to do, who is to do it, where it is to be done, when it is to be done, and tell your people why, when they need to be told why (which is the case far more often than not). Continue supervising the job until it has been done better than the person who wanted it done in the first place ever thought it could be.

6. Know your people and look after their welfare Good leaders always get the best they can for their people by honest means. With judgment, you'll know which of your people is capable of doing the best job in a particular assignment. Leaders share the problems of their people, but they don't pry when an individual wants privacy.

7. Everyone should be kept informed Make sure your subordinates get the word. Be known as the one with the honest, complete answers. Squelch rumors. They can create disappointment when they're good but untrue. They can sap morale when they exaggerate the problems at hand. Have the integrity and the dependability to keep your people correctly informed.

8. Set goals you can reach Don't send two people to do something that calls for five. Your people are good, but don't ask the impossible. Know the limitations of your resources and bite off what you can chew. Those who have a reasonable goal and then achieve it are a proud lot. Next time, they'll be able to tackle a little more.

9. Make sound and timely decisions Knowledge and judgment are required to produce a sound decision. Include some initiative and the decision will be a timely one. Use your initiative and make your decisions in time to meet the problems that are coming. If you find you've made a bum decision, have the courage to change it before the damage is done. But don't change direction any more than you absolutely have to.

10. Know your job This requires no elaboration. It does require hard work on your part. Stay abreast of changes. Read up on recent developments. Don't be the one who always talks about "The way we used to do it."

11. Teamwork Train your people as a unit. Put your people in the jobs they do best, then rotate them from time to time. They'll learn to appreciate the other person's task as well. When one member of your team is missing, others can do their share. When you and your people have done something well, talk it up. This builds esprit de corps. You can't see it but you can feel it. An organization with a lot of esprit holds itself in very high regard while sort of tolerating others. There's nothing wrong with that. Everyone has a right to believe that their department is the best in the whole organization. After all, they're in it!

These eleven principles are not project oriented or profit oriented: they are people oriented. Without conscious, constant effort to care for and understand people, including oneself, everything else a leader does is drastically devalued. General of the Army Omar Bradley once stated flatly that, "The greatest leader in the world could never win a campaign unless he understood the men he had to lead."

These people-oriented values are considered mainstream in the military. How mainstream? These authors once wrote an article titled, "Love and Leadership" proposing that the relationship between a leader and the led must include much of what is best about a strong and lasting love between individuals: mutual trust, a constantly evolving relationship, a generous sharing of triumphs, and a deep and abiding concern for the well-being and future of the other. It won an award from the MCCCA, the Marine Corps Combat Correspondents' Association, as crusty a bunch of veterans as you're likely to find.

You can tell from the stories that endure in the military what people truly believe. Maybe that's why one that gets told over and over in the Marine Corps, with great glee, concerns three lieutenants. Not coincidentally, it's a story about leadership styles—although it has the grace not to belabor the point.

It seems that there were these three young lieutenants who reported in to their first duty station at the same time. The personnel officer knew that he had three jobs available: one job was just the thing to launch a brilliant career, one job was a good-enough job, and one job was a sure kiss-of-death. The problem was how to determine which lieutenant received which assignment.

He decided to pose a problem. He said to the three, "I need to raise a flagpole. This is a list of the equipment that will be available to you, and I will give you my best sergeant and four good troops. I want each of you to think about it for an hour and then tell me how you would get the job done."

After an hour, one lieutenant returned and said, "I have figured it out completely. Here—look at my calculations. I've calculated exactly how deep to dig the hole, the angles for everything—every detail. I would show these calculations to the troops, answer all questions, and then we'd have the flagpole up in no time at all."

The next lieutenant came in and said, "Well, sir, I've never put up a flagpole before so I would sit down with the sergeant and the four good troops and say, 'We need to get a flagpole up. Here is a list of equipment available to us. What are your suggestions?' I would listen to them, ask questions, let them ask each other questions and build on each other's ideas—and then I would make the decision. I would announce my decision, explain my decision, and answer all questions. Then we'd put up the flagpole."

The third of the lieutenants walked in and said, "I would walk over to the sergeant, hand him the list of equipment, and say, 'Sergeant, put up the flagpole.'"

Any organization that rewards leaders with that kind of courage, that willingness to take a personal risk, is half way home to quality.

The Quality Observer, *July 1992.*

THE THREE PRIORITIES OF LEADERSHIP: LESSONS FROM THE MILITARY

With traditional top-down leadership practices proving ill suited for our times, you might think the military would be the last place to find appropriate models of leadership. Think again. The fact is, for over 2,500 years, the military has developed a set of leadership guidelines and characteristics that are both humane and effective.

How can the ultimate command-and-control organization make such a claim? Precisely because leadership in the military is viewed as a behavior, not a position. There is no question that if two people are present, the senior one is in charge. There is also no question that *every member of service has to be ready to act in a leadership role.* This is especially true in the armed services of the United States.

On the 50th anniversary of the invasion at Normandy, several commentators marveled at how the loss of the titular leader of a unit did nothing to slow the attack. As each man fell, the next-senior man stepped forward and assumed command. Decimated units reformed under men with little rank and less experience and fought their way off the beach, using routes and methods never dreamed of by the planners of the assualt. Meanwhile, German units, constrained by the Nazi hierarchy and incapable of the same flexibility, were overwhelmed.

The military has always used what the quality movement now calls self-directed work teams. Working under broad directives, military units are left to their own devices at every level of the chain of command. From a team of recruiters in a town far from headquarters to a behind-enemy-lines special combat team to the captain (often a junior officer fresh out of the Coast Guard Academy) and crew of a cutter, small units are expected to make independent decisions with very real impact.

Leadership training in the US military begins on the first day of enlistment and never ends, and the basics of leadership are consistent. By teaching the same principles to all, the armed services allow for conversation and shared learning between the ranks. Implications of a teaching point are different for a corporal than for a colonel, but the same concepts apply and each rank addresses the same points.

For example, everyone in the military understands and shares the same three leadership priorities. The first is to accomplish the mission; the second, to take care of personnel; and the third, to create new leaders. Can the same be said about a civilian organization? Does everyone share not only the commitment to the mission but the commitment to their colleagues? Ask most civilians about their second priority at work and the response is likely to concern personal career enhancement. Military people are not angels; they are as concerned with their careers as anyone else. They know, however, that advancement is a by-product of success in meeting leadership priorities, not a goal in itself.

To teach leadership skills, the military uses lists to organize information. One such list is the US Army's 11 Leadership Principles (see box belows). Lists are supported by a rich and varied curriculum that draws from vast experience under difficult circumstances. Among the points routinely debated and refined are the value of initiative and the true meaning of discipline (including the willingness to act appropriately in the absence of orders).

US Army Principles of Leadership

- Know yourself and seek self-improvement.
- Be technically and tactically proficient.
- Seek responsibility and take responsibility for your actions.
- Make sound and timely decisions.
- Set the example.
- Know your soldiers and look out for their well-being.

- Develop a sense of responsibility in your subordinates.
- Ensure the task is understood, supervised, and accomplished.
- Build the team.
- Employ your unit in accordance with its capabilities.

Despite the stereotypes, blind obedience and abuse of power are anathema to members of the military. The outrage over the recent sexual harassment scandal in the Army was shared by service personnel. In an article entitled "The Enemy Within," the *Army Times* reported on actions taken to remedy the situation—criminal charges, replacement of the company commander with a black woman captain, and establishment of a hot line to determine the extent of the problem. More than a few corporations could take lessons.

And disobedience to orders is not unknown. During the Korean War, Gen. Oliver P. Smith, USMC, refused an order by Gen. Douglas A. MacArthur, USA, which would have dangerously exposed his troops in the rush to the North Korean border. General Smith was subsequently vindicated. His behavior reflected a number of leadership principles on the list: He was tactically and technically proficient, he knew his men and their capabilities, he made a judgment call as to what constituted their breaking point, and he protected subordinates at the risk of his career.

"I was just following orders" is never an adequate defense. Members of the military are expected to develop a moral compass: *who* you are is considered just as important as *what* you do. The Marine Corps' list of 14 Leadership Traits identifies desirable characteristics on which to base internal controls. External controls lead to situational ethics; internal controls lead to responsible behavior. The field manual *Leading Marines* states, "If you are prepared to talk about your actions, or lack thereof, in front of a national audience, made up of all your seniors, peers, subordinates, and friends who share the same professional values, and whose opinions you value, then your behavior was, or is, probably ethical in nature."

Which is not to say that following orders is not important or that authoritarian orders are not given. The military teaches three leadership styles: authoritarian, participative, and delegative. But not all styles are created equal. Authoritarian leadership is only acceptable as long as three conditions apply: you are out of time, you have all the information, and your people have high morale. High morale is the result of using participative and delegative leadership styles whenever possible. Troops (as well as employees) know that if everything is an emergency, then nothing is an emergency—and respond accordingly.

Surprisingly, participative leadership does not enjoy the same status as delegative leadership in the military. One training example makes this point. Each of three young lieutenants is given an experienced sergeant, four good soldiers, equipment, a deadline, and a problem: how to put up a flagpole. The first lieutenant struggles with calculations, calls the team together, and issues orders; the second asks the sergeant and soldiers for input, makes a decision, and issues orders; the third turns to the sergeant and says, "Put up the flagpole. If you need me, I'll be available." The military holds up the last as the ideal. Delegative leaders develop a sense of responsibility in subordinates (see "US Army Principles of Leadership"), create new leaders (one of the three priorities of leadership), and bind together the leader and led in a covenant of mutual trust. There is no better way to show confidence in others than to delegate authority equal to responsibility.

One point is never in doubt in military leadership theory: Leadership is a subset of love. In an interview, Gunnery Sgt. William E. Hazelwood, an instructor at the Marine Corps Recruit Depot Drill Instructor School, stated simply, "You do realize, sir, that leadership is all about caring." Leadership and love share a blend of rational and emotional elements that make them inseparable, and it is difficult not to conclude that if a person is not good at one—love or leadership—he or she is probably not very good at the other.

US Marine Corps Leadership Traits

• Integrity	• Justice
• Knowledge	• Enthusiasm
• Courage	• Bearing
• Decisiveness	• Endurance
• Dependability	• Unselfishness
• Initiative	• Loyalty
• Tact	• Judgment

Love is demonstrated in a number of ways. Love is at the core of the second leadership priority: take care of personnel. Field Marshal Lord Slim, to drive home the inter- connectedness of all his units during World War II, put his staff on half-rations whenever his forward formations had to go on half-rations. Battlefield examples of love abound. A story about Gen. Joshua Lawrence Chamber- lian, a famous Civil War hero, demonstrates both the intensity and the tenderness of the relationship possible between leader and led.

Chamberlain never put his men in harm's way without going there himself. His valor and regard for his troops drove some of his own sergeants to take extraordinary risks to protect him during battle, as historian Alice Rains Trulock recounts: "Rushing into the open, they seized him bodily and carried him to the shelter of the works, risking their lives and possible discipline for such insubordinate action, but demonstrating their affection for him in a way that could not be doubted, only pardoned."

General Chamberlain showed his love by his example and his love was returned in full. What corporate leader would not want to inspire the same response?

Leader to Leader, *Spring 1997.*

WHAT THE MILITARY CAN TEACH BUSINESS ABOUT LEADERSHIP

United States business needs leadership. The distinction between leadership and management is becoming ever clearer as more books and articles are published on this topic.

Yet the journalistic–academic complex, which plays the lead role in defining what information is presented to American executives and executives-in-training, stubbornly refuses to draw from the experiences of the only segment of society that has debated and studied leadership for 2,500 years: the military.

What can the military teach business executives about leadership? Many lessons about adapting style, applying values and ethics, training subordinates and successors, and loving the people you lead.

STYLES OF LEADERSHIP

First, the military codifies the issues in a simple, realistic way. The US Army Leadership Manual (FM 22–100) defines three styles of leadership – authoritarian, participative, and delegative.

Authoritarian. The first is described in pragmatic terms: "A leader is using the authoritarian leadership style when he tells his subordinates what we wants done, and how he wants it done, without getting their advice or ideas. Under the following conditions, the authoritarian approach is normally appropriate:

- You have all of the information to solve the problem
- You are short on time
- Your subordinates are motivated

Sometimes people think that a leader is using the authoritarian style when he yells, uses demeaning language, and leads by threats and abuse of power. This is not the authoritarian style. It is simply an abusive, unprofessional style of leadership."

The manual then explains that the way a leader insures that he has all possible time and information and his people are well motivated is by practicing participative and delegative leadership most of the time.

Participative. This style involves "one or more subordinates in determining what to do and how to do it." In the case of participative leadership, "the leader maintains final decision-making authority."

Delegative. Using this style, "the leader delegates decision-making authority to a subordinate or group of subordinates." The leader "is still responsible for the results of his subordinates' decisions." It is an oft-repeated rule in military leadership that, "You can delegate authority but not responsibility."

APPLYING VALUES AND ETHICS

Second, the military emphasizes the character ethic over the personality ethic of leadership. For example, chapter headings in FM 22–100 include the following:

- "Professional Beliefs, Values, and Ethics"
- "The Character of a Leader"
- "Leadership That Provides Direction"
- "Principles of Leadership"
- "Know Yourself, Seek Self-improvement"
- "Set the Example"
- "Keep Your People Informed"

Focus on accomplishing the mission and caring for, and about, the people assigned to a leader are the common threads that run throughout the military handbook. This focus becomes a spring-board for discussions of the difference between management and leadership in other military sources, a debate which is concurrently taking place in the civilian community. In the *Marine Corps Gazette*, the difference is defined as, "A manager cares that the job gets done, a leader cares that the job gets done and he or she cares about the people who do the job."

PRINCIPLES OF LOVE

In an article which received a commendation from the Marine Corps Combat Correspondents' Association, a Marine officer discussed leadership as being a subset of love, adding that if a person wasn't capable of love, he or she would most likely not be much good at leadership.

There are many varieties of love and one of the greatest is good leadership. Since leadership is a form of love, our knowledge and experience of the root concept can provide us with useful insights into leadership. Perhaps the most obvious is the act of caring for the welfare—the well-being, physical, and mental—of others.

A person who would call himself or herself a leader of Marines must be capable of loving and being loved. To love someone is to make a commitment to him or her, a promise to work hard to improve conditions for them and to improve yourself. It is not a pledge to nag them until they finally shape up, but rather a promise to work with them toward mutual goals, a higher state.

A platoon commander, while laying the groundwork for the reputation that will follow him throughout his career, must truly love his men if he wishes to be known as a good leader. If he brings technical expertise and ambition but no warmth to his position, his troops will return his investment in like currency. They will do precisely what he says, but will not give the extra effort that is the mark of the well led. Neither will they overlook his faults nor compensate for his mistakes.

ON-GOING TRAINING

What of all the stories that everyone has heard about sadistic Drill Instructors, power-mad officers, and devious, cruel sergeants? Stereotypical beliefs about how the military gets from here to there abound. For many, the words "autocratic" and "military" are interchangeable as adjectives in front of the word "management."

The stereotype was not manufactured out of thin air. There have been—and continue to be—enough incompetents in the uniforms of every country to sustain the unflattering, albeit generally inaccurate, image.

But the fact remains that the stereotype does not reflect reality. The military has been compelled to trust its people. In combat, any person can become a casualty at any time; replacements must always be ready to step up, and in. Subordinates have to be told what is going on, and they have to be prepared to assume responsibility.

"Succession planning" is automatic—and very open. The military is made up of people all busily training in some way to do something that they pray quietly at night never happens.

The peacetime version of this same phenomena is encouraged by the constant transfer of military personnel. It is rare to be in one geographical location for more than three years, rarer still to be in the same job that long. The training of one's subordinates is a never-ending task.

The penalty for failure to involve, and insure the growth of, subordinates is severe in the military. If a business manager chooses to be autocratic, to not enlist or use the input of any of the people who work with or for him or her, the worst thing that can happen is the inclusion of a pink slip in the next paycheck—devastating, but not fatal.

On the other hand, if a military leader in combat fails to draw on the knowledge, experience and intuition of subordinates, he could die as a result. Most unfair, he could take others with him. This potential for disaster makes choosing to build teams that maximize everyone's strengths, and to be participative and caring, a logical imperative.

Principles of Leadership

1. Take responsibility for your actions and the actions of your people; handle responsibility with judgment, tact.
2. Know yourself; seek self-improvement and objective evaluation; be honest with yourself about yourself.
3. Set the example.
4. Develop your subordinates.
5. Ensure that a job is understood and then supervise and carry it through to completion.
6. Know your people and look out for their welfare; share problems; don't pry.
7. Keep every person informed.
8. Set goals that stretch but are within reach.
9. Make sound and timely decisions.
10. Know your job.
11. Promote teamwork.

Principles of Love

1. Take responsibility for your actions; share responsibility at appropriate levels of maturity and initiative; treat loved ones with judgment and tact.
2. Know yourself; seek personal improvement, don't lie to yourself or to them.
3. Don't make demands that you wouldn't want made of you.
4. Serve as a resource in your loved one's growth and invite them to be part of yours.
5. Seek to know your loved one's needs and wants and show concern for their physical and mental welfare; be available to listen as they share problems.

6. Make sure that any important requests you make are correctly understood.
7. If something is bothering you or affecting your performance, tell your loved one about it tactfully.
8. Manage your expectations; you're not going to be in Shangri-la every day.
9. Make necessary decisions as well and responsively as you can.
10. Build and nurture solid, mature relationships.
11. Share with and work with your loved ones toward common goals.

William F. Ward, commanding general of the 77th Army Reserve Command and president of Gestam, Inc., was quoted in *Success* as saying:

The military is more people-oriented than business. My experience with business schools is that they teach a lot about the skills of management but very little about the qualities of leadership. And it shows. I've seen hundreds of companies where the CEO didn't care about people. He drove out all the good people. It's strange because the military is perceived as being this great faceless group that shouts and screams. But at the managerial levels in the military, employees have more to say about how they do their jobs than in most businesses.

Business leaders can learn from military experience. The popular perception of the military is improving. The military is now held in high regard in public surveys (considerably ahead of "big business" and even further ahead of Congress). Couple this with the wide availability of written materials on military leadership—and a valuable source of information is waiting to be discovered.

Executive Excellence, *Feb. 1989.*

10

THE PAUL REVERE PROCESS

In 1985, Tom Peters named the "Quality Has Value" process the "best quality process in any service organization in North America." Even more than a decade after the Paul Revere process was defined and initiated, it remained one of the leading examples of 100% participation in quality for service or manufacturing organizations. The process is detailed in the author's work *Commit to Quality*.

Written at four different points during the evolution of the Paul Revere Insurance Group's process, the four articles in this chapter overlap—but the story is worth telling and retelling. On the strength of improvements from "Quality Has Value," the company became the national leader in its field and was a finalist for the Baldrige.

Alas! This is another story with an unhappy ending. Content to rest on its laurels after five successful years, the process languished. The company subsequently was acquired by a larger company and the process was discontinued by the new management team.

INSURANCE FIRM SHOWS THAT QUALITY HAS VALUE

"Quality in Fact" and "Quality in Perception." With manufactured goods, it is customary to think of one following the other. A measurable improvement in the first can, with considerable rapidity, change the second. Witness Chrysler.

In a service industry, however, the relationship is not as clear. The two concepts are more closely intertwined, and it is often difficult to determine where one leaves off and the other begins. Consider a phone conversation between a clerk in an insurance company home office and an independent agent or a policyholder. Is the level of courtesy, accuracy, and promptness that the clerk exhibits contributing to quality in fact or in perception? And in instituting change, where is the most effective place to begin?

When designing a quality process in a service industry, managers must address both the blurring of the line between fact and perception, and the level of personal contact between customers and employees other than salespeople. A quality process at The Paul Revere Life Insurance Companies was created to address both of those needs.

The entire process became known as "Quality Has Value" (QHV). We took it so seriously that we changed our slogan to "Our Policy is Quality" – which now can be found on practically every piece of paper that leaves company headquarters in Worcester, Mass.

In the first 15 months of this process, our employees generated over 10,200 new ideas for improving quality; 6,100 were implemented. (Figure 10.1 charts this progress through the end of our fiscal year). Total savings for these ideas were $4,870,000 (Figure 10.2 shows the total savings for our fiscal year ending Nov. 30, 1984)— with hard dollar savings topping $1,500,000. In addition we made an exhaustive study of the operations carried out in our company, department by department—leading to significant increases in efficiency. These results were due to a combination of good planning, unwavering support from top mangement, and more than a little luck.

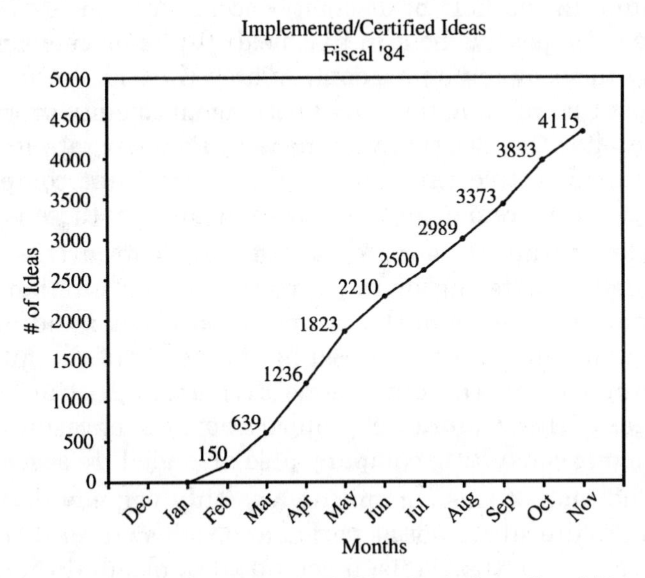

Figure 10.1 Implemented/Certified ideas fiscal '84

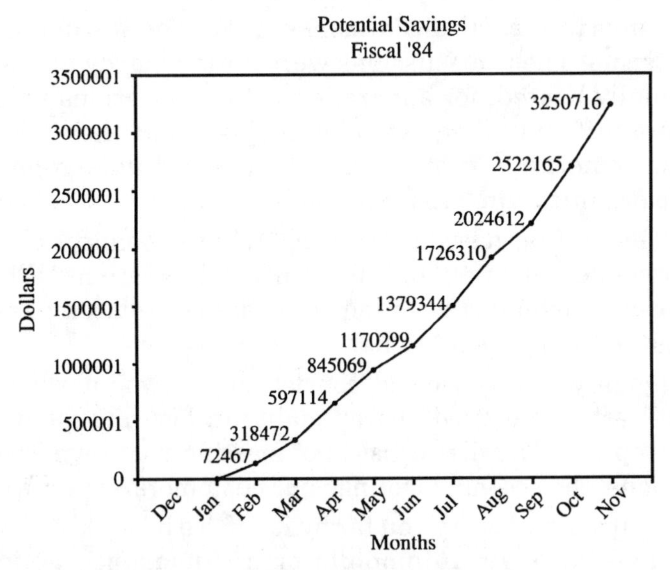

Figure 10.2 Potential savings fiscal '84

The planning stages of QHV started in 1983, after Paul Revere—which for many years had comfortably led its competitors in the field of disability income insurance—began to feel competitive heat in that field. Top management, notably the company's president, Aubrey K. Reid, Jr., decided to institute an ambitious, and permanent, quality program. Soon after this decision was made at the corporate level, a similar directive came down from our parent company, AVCO Corp., reinforcing our resolve to address these issues.

At this point, a quality steering committee was formed, consisting of the number-one or number-two executives of each of the company's divisions and major departments, and co-chaired by the heads of the human resources and the operations divisions. (In the latter stages of this committee's deliberations, a representative from our consulting company also attended its sessions.)

The task of this committee was only vaguely defined, and the literature not as varied and rich as it has become in recent months. Thus, a good deal of original thought, and not a little good luck, came into play.

Important early discussions centered on the definition of terms. These discussions were often lengthy and occasionally heated, for apparently simple questions become more difficult when everyone realizes that they are setting something in place that is supposed to be a part of the organization "from now on."

The first discussion question, "What is quality?"evoked a two-part answer, and an additional question, "Who is our/my customer?"resulted in a far more encompassing answer than is traditional.

Quality, it was decided, consists of the two interdependent parts mentioned above: quality in fact and quality in perception. The first consists of meeting your own specifications; the second, meeting your customers' expectations.

Neither will carry you far by itself. To be making something—be it an automobile or an insurance policy—

exactly as you intend to will be to no avail if your potential market believes you are producing an inferior product. On the other hand, customers may come flocking if they believe you to be a quality producer (thanks perhaps to old memories or a new salesman), but if your product doesn't measure up to expectations, not only will they not come back, they'll tell their friends.

Defining a customer was an equally complex task. A customer is not *just* the money-waving "ultimate" consumer. A customer is anyone to whom you or your work unit provide products, service, or information. In our case, it wasn't only the policy-holders about whom we wanted our employees to be concerned; it was also the people in the next department, or perhaps at the next desk. The same principles of quality in fact and quality in perception apply, regardless of which customer is being considered. If the person at the next desk believes the quality of your output to be low, then that must be addressed. Either the quality of your output will have to be improved, or your customers must be made to understand why those expectations cannot be met.

Having decided on these matters, the committee went on to the specific techniques through which these ideas would be implemented. In the end, we decided on four techniques. These were:

- *Quality teams*—much like the better known quality circles, except that membership in one or more quality teams was compulsory for all employees.
- *Value analysis*—through which we thoroughly examined the procedures in each department of the company.
- *Surveys*
- *Changes in corporate culture*—through which the idea of quality could take permanent root in our company.

These four procedures—and especially the first two—were the framework on which we built the QHV program. I will describe each of them here, but the reader must remember that the procedures outlined here were partly the result of planning by the steering committee, and partly of evolution.

In fact, any company that wishes to institute an ambitious program like ours had best remember that once you have started, you can turn back or slow down only at your peril. Once you let the employees in on the responsibility and authority, they will greatly begrudge ever having to return it. As a result, you'll soon be a different company. If you do the job right, you'll also be a richer one. Either way, though, the process takes on its own momentum, and therefore must be planned with care and caution.

Quality teams These teams, like traditional quality circles, were made up of about ten employees each, and met about once a week. In two important ways, however, they differed from quality circles. First, membership in a quality team, and attendance at team meetings, was compulsory. Second, the teams had more power than circles—and so could actually implement their ideas without prior approval from management.

These teams were instituted throughout the company, at all levels. Each team was generally headed by the "natural" boss of the work unit the team represented, but this was not mandatory. Team leaders at one level of the corporate ladder would be members of the quality team at the next higher corporate level—the team, most likely, that was led by his or her boss.

The job of the teams was to come up with what we called quality ideas for improving the work process—ideas for changes not only in each team's own work area, but throughout the company. In its own area, a team could implement its own quality ideas. The effectiveness of these ideas would then be checked out by special quality personnel—and if they passed muster, would become part of the permanent company procedures.

To help control this whole process—and to keep it from becoming too complex—we took a couple of important steps. The first of these was instituting a training program for quality team leaders, to help them direct the team meetings. This was especially important, because these meetings were to last only one-half hour, once a week—and would therefore have to be wonderfully efficient.

After a brief attempt at constructing our own course, we learned of the Zenger Miller course, "Group Action." This course—a 40-hour program in which the goal is to train people to conduct meetings in a participative manner, drawing out ideas and problem-solving steps from others—fit our purposes well. We designated several employees, myself included, to be instructors.

Our second control step was to determine how to keep track of what each team was accomplishing. Since I was the newest arrival, having joined the company about the time the team leader training began, and had the fewest defined duties, much of the definition of the "mechanics" fell to me.

During the closing months of 1983, the quality team tracking program (QTTP) was developed. It made the process virtually paper-free and thus team-leader-friendly. It also took advantage of an idiosyncracy of many service companies, i.e., the abundance of computers.

The QTTP is a big file; each team leader is able to alter only his or her piece of the file. Anyone can look through the entire file, including the company president's team file. The four key pieces of information on the screen are: team leader name, idea description, idea status, and estimated savings.

Whenever a team picks an idea or a series of ideas to work on, these can be added to the file—one idea per screen—with an "idea status" of "1." These screens can be changed and updated. If an idea is implemented, either independently or after calling in appropriate help, the status is changed to a "4."

The QTTP became the link between the quality teams and Quality Team Central. It became the single most important tool in implementing the quality team aspect of QHV.

Paul Revere had had a work measurement program in place for several years, as well as work simplification projects and a suggestion program. Four productivity analysts had directed all three operations. They knew the workings of the individuals and sections, and the interrelationships among departments, better than any other group in the company.

Since, I argued, the QHV process would make all three of their functions obsolete, these four people were natural candidates for the formation of Quality Team Central. The availability of the perfect people for Quality Team Central was one of several instances of good luck that have made our program work.

Each Friday, Quality Team Central would get a printout of all the 4s on the system, i.e., all the newly implemented ideas. Note that it is *implemented* ideas that the Quality Team Central analysts react to. This marks a radical departure from the standard quality circle approach, with its presentations to management. The teams are being trusted to know what they can implement and what isn't fully within their area; what affects other departments, and thus requires their cooperation, and what doesn't. In short, they have the responsibility to determine ways of improving our quality and the authority to do something about it.

The control element came in the Quality Team Central analyst's after-the-fact check. The vocabulary chosen was that a team implemented an idea while Quality Team Central certified the implementation. And only certified ideas counted for recognition purposes.

Certification consisted simply of the team leader showing or explaining the implemented idea to the analyst, whose knowledge of the company enabled him or her to spot cases when some necessary coordination or notification had not taken place. The analyst would also double-check the team leader's figures if there were any savings resulting from the implementation of the idea. Once an idea was certified, the idea status was changed to a "5."

Other parts of the quality team process were the PEET program, the recognition system, and the definition of "Quality News." PEET is an acronym for "Program for Ensuring that Everybody's Thanked." It is a formalized approach to "management by wandering around," a concept made famous by Tom Peters and Bob Waterman. Each Monday, President Reid and the six executives who report directly to him receive a PEET sheet. On it are the names of two quality team leaders, along with a description of their teams' activities to date. It is the executive's responsibility to make the time during the week to seek out these two leaders and talk with them.

One of the goals of the PEET program is to set an example for middle managers. The hope is that after middle managers have seen top executives come into their areas and sit and talk with various supervisors and managers, they will decide, if only out of self-defense, that talking with people is a good investment of time. Also, since most of the middle managers are themselves team leaders, it provides an unstructured forum for them to periodically talk with each of the top executives.

Throughout the first year, middle managers were, in fact, the slowest group to adjust to the quality process. All their corporate lives they had been taught that wisdom came from further up the corporate ladder; you listened up and proclaimed down. Now they were being told to "listen down"; authentic change could (and did) initiate at the lowest organizational levels. Many found it uncomfortable, if not threatening. Fortunately, time (and success) helped change their perspective.

As for recognition, we replaced our "suggestion" program—which paid an employee 25% of the savings to the company for an idea suggested by him or her—with pins,and prizes from catalogs for all the members of teams that reached certain goals. Each member of a team that reached ten implemented ideas, or $10,000 in savings to the company, would get a bronze pin. At 25 ideas or $25,000 in savings, it would be a silver pin and a gift from a catalog. At 50, the teams got gold pins—awarded by the company president—and prizes from a more expensive catalog.

Had we been fully aware of the average number of completed ideas per year by a quality circle in the US (six versus between four and five in Japan), we never would have set the standards so high. As it was, we expected to see our first gold teams in about September or October, given a kick-off in mid-January. In fact, teams started making it to gold on the strength of 50 implemented ideas in March. And by year end, the average team completed 37 ideas. (See Figure 10.3)

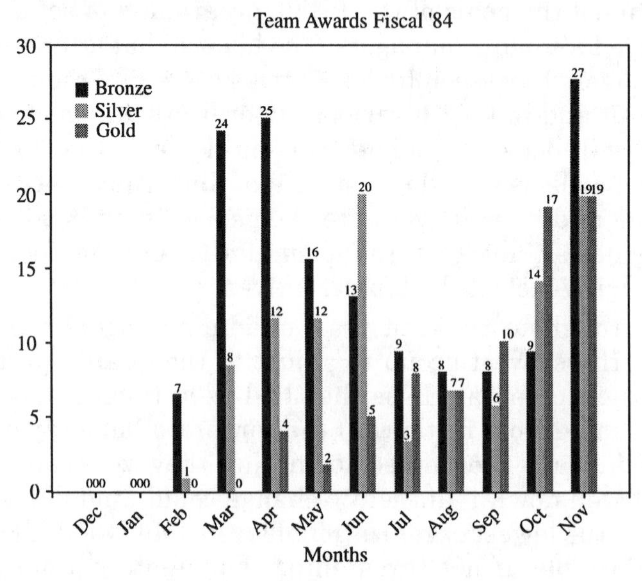

Figure 10.3 Team Awards Fiscal '84

A final organizational detail was the creation of a vehicle for communicating results of the QHV process to all employees. A newsletter, "Quality News," was designed to fit this need, with initial publication set for March 1.

When we finally inaugurated the QHV program in January, 1984, after many months of planning and training, employees had heard enough about it for many of them to become convinced that this wasn't just "this year's program" that would soon go away. Not everyone "bought in" immediately, of course. But enough did to get it rolling.

The non-voluntary nature of our quality teams compensated nicely for the natural hesitation of many people to enlist at the beginning of any new project. In quality circles, a late-coming believer must go through a fair bit of paperwork and time to form a new circle. Our late converts simply had to open their mouths: they were already functionally "in"; they had already been attending meetings.

Having a staggered acceptance rate even has its advantages. About the time the early participants are beginning to slow down, the others, seeing all the ideas being implemented all around them, start contributing.

Two specific areas keyed the early spurt of ideas. One was the keypunch section. Our company still uses "old-fashioned" computer cards for some of its input. The workers in keypunch were a classic case of people being treated as if they were incapable of thought. No one had asked them in recent memory what they thought; in fact, they had been told not to think, but rather to type what they saw. They were inputting errors—and they knew it. They were using inefficient formats—and they knew it. Their section supervisor and quality team leader believed what she'd been told about the QHV process and she convinced her people.

The ideas flooded out. The keypunch operators surprised other employees by literally refusing to be partners in low-quality work. If they knew something was wrong, they wouldn't just type it, knowing they were going to see it again in two to four days. Their team name was "Goldfingers," and they were the first to the silver level.

The other early hotbed of ideas was the financial services division. The catalyst there was a level-three clerk, new to the company, who couldn't figure out why the company didn't maintain a minimum balance in its various checking accounts (used to pay claims, etc.) around the country. Our goal was a zero balance—a practice no doubt instituted before interest-bearing NOW accounts, high charges for overdrafts, and free service for maintaining a designated balance.

The clerk figured out that the practice was costing the company about $3,000 per year. She and her team leader went to the division vice president, who said the only thing wrong was that the figures were too conservative. The cost was more like $5,000. The policy was changed on the spot, the money shifted the next day, and the signal sent—QHV was for real.

In the following months, quality ideas varied widely in their content and effect. Many, particularly at first, were small. They consisted of changes in forms or in the sequence of steps taken to accomplish a particular task. As the year progressed and the teams gained more and more confidence in the support and trust being offered, ideas became more complex. Team leaders also became more aware of the value of their time and that of their team.

The biggest single idea (in terms of monetary value) involved the reissuing/resolicitation of a group policy after some changes in phrases. The changes were originally put in place to avoid double payments or expensive lawsuits. The rewrite made the policy clearer and in fact allowed us to lower the premiums.

The quality team involved decided that rather than simply changing the policy, they would brag about the improvement to the current customers and reenroll everyone. It meant extra work for a short period of time, and a bit of risk, but it paid off in $193,000-per-year increase in premiums paid in.

Value analysis The second major effort—in addition to quality teams—was our value analysis workshops. In these, the decision makers—from supervisors to vice presidents—were offered the opportunity to stand back and take a look at their work units. This was true for everyone from underwriting to food services, from human resources to building maintenance/engineering.

The problem was to first decide what the primary objective of each group was and to then determine how it could best be fulfilled. A useful gimmick was to require that the primary objective and the secondary supporting goals be described in just two words—one verb and one noun.

The value analysis workshops were not the blood-letting that is sometimes associated with consultant-driven value analysis programs. For one thing, the promise had been made that nobody would be removed from the payroll of Paul Revere because of the QHV process. Your particular job might disappear, but if it did, you would be offered another. Expected decreases in complement would be taken up by natural attrition.

Once the workshop—with the help of some pointed questions/suggestions by the facilitator—defined its goals and then listed its various responsibilities, participants were given the opportunity to blue-sky for a bit. After this exercise—which recognized no monetary or personnel constraints—was concluded, the real world was reintroduced and the list of ideas was critically reviewed by the workshop. By working in this sequence, many ideas that might have appeared a bit bizarre at first could be shaped into fruitful ideas or serve as catalysts for a practical idea that might otherwise have been overlooked.

The resulting list of recommendations was written up by the facilitator, but it was signed by the workshop participants. It was *their* list of recommendations on how to improve the quality of their unit—not those of a consultant. This increased the chances of implementation manyfold since the originators and the implementors were virtually the same people.

During the first 12 months, about 30 workshops were conducted. The potential annual savings of all the recommendations is over $6,000,000.

The interplay between the quality teams and the workshops proved to be mutually beneficial. By the time a group began a workshop, they had grown used to discussing new ways of doing things. They knew that having an idea about changing things was not an admission of having done things wrong.

Similarly, when the workshop was over, the habit of being creative and looking for better ways was stronger. The original decision to proceed with both the quality teams and the value analysis program at the same time had been the right one.

As the months went past, we began to introduce some short-term programs to help guard against the expected letdown. As is evident in Figure 1, the letdown never really came. While some teams would slow down and not be heard from for a month or two, the overall flow of ideas remained remarkably constant.

Surveys Surveys became a normal event during 1984. The first one, conducted in February, had the most direct effect on the QHV process. We required all quality teams to compute their cost of quality. Philip Crosby breaks down cost of quality (COQ) into three components: prevention, detection, and correction. We divided the third component into correction and correction-failure. The split between correction and correction-failure was determined by whether or not the business affected by the error was retained, or if the cost represented a permanent loss.

The calculations drove home to everyone that we were talking about a big ticket item, since the total cost of quality ran well into the millions. They also helped to focus teams quickly on areas that needed improvement. Whenever a team lagged during the year, it could always return to its COQ calculation for "inspiration."

More traditional surveys were conducted with our home office employees ("What do you think of your quality? Of other departments? Of the field force?"), with the field/sales people ("How's the home office doing...department by department?"), and with our policy-holders ("How are we doing?"). These confirmed what was suspected—we were pretty good, but there was definite room for improvement. It also identified specific areas to be addressed.

Corporate culture To help strengthen the slowly emerging corporate culture change, the top executives began the Tom Peters–Zenger Miller course, "Toward Excellence." The program consists of 15 videotape segments, each approximately 15 minutes in length. The tapes consist of Peters talking with various people about the points laid out in *In Search of Excellence.* These points have been developed in subsequent research.

Each segment is supported by a little advance reading after the tape, evaluation, discussion, and specific action-step planning take place. While the tapes and exercises can be "gotten through" in three days or so, the decision was made to extend the process over several months. This way, the executives had time to assimilate each step into their personal "bag of tricks" before looking at the next one.

A major philosophical point, of course, is the difference between a program and a process. A program is perceived as something with a predetermined life span. Many short programs were used, and will be used, to supplement and support QHV. But QHV itself is a process. It is the way we will do things from now on—a corporate way of life.

Looking back over our first year, QHV '84 definitely sent us off in the right direction. No specific focus was given to the quality process in general or to the quality teams other than to improve the quality in fact and perception of Paul Revere. Most teams directed their efforts inward. They looked for ways to make their own jobs easier, or more efficient, or more enjoyable—and they made lasting improvements. The numbers are impressive.

In QHV '85, we have defined a two-fold focus—the customer, and the cost of nonconformance. Ideas that have a direct, positive effect on any of a quality team's many customers will receive extra credit toward earning recognition. Ideas that help avoid nonconformance—doing it wrong and having to correct it—will also receive bonus credit.

The recognition program has been altered and the search is on for short-term programs to encourage and thank our employees.

Again, the bottom line of QHV remains the change in corporate culture. That has begun. It is evident in the fact that "quality" has become a natural word in people's vocabulary; it's evident in the continued strength of the QHV process; it's evident in the coming aboard of even die-hard skeptics.

Perhaps it was best exemplified by a bouquet of flowers that a clerk in the policy holder services department received one day late in 1984 from a field agent. The note said, "Thanks for the Quality job."

Quality Progress, *June 1985.*

REFERENCES

1. Thomas Peters and Robert Waterman, *In Search of Excellence* (New York: Harper and Row, 1982).
2. Philip B. Crosby, *Quality is Free* (New York: McGraw-Hill, 1979).

THE POLICY IS QUALITY

George Odiorne, a renowned consultant, has stated that white collar productivity will be the hottest management topic of 1985. It is a problem that the Japanese, famed for their success in increasing the productivity of manufacturing firms, have only begun to study.

One American conglomerate, the Paul Revere Insurance Companies, has put together a two-pronged quality process which, after one year in full operation, appears to have made strides toward solving the puzzle of how to increase both the quality and the productivity of the white collar industry.

By adapting and blending several techniques, and adding a few original touches, the Quality Has Value program reaped results. Between January and November of last year, 4135 ideas originated by employees on quality teams at all levels were implemented, with annualized savings in excess of $3.2 million. The other prong of the dual approach, Value Analysis, generated ideas that, when fully implemented during the coming months, are expected to account for another $6 million in annualized savings. The improvements that have come out of these two parts of the process range from mundane administrative procedures to complex, insurance-particular ideas.

There are four components of the Quality Has Value program—value analysis workshops, quality teams, surveys, and a corporate culture change. The first two are the instruments of employee involvement and change; the third is used both as a source of ideas and a measure of progress; and the last is expected to result, over time, from the first three.

The quality teams bear a superficial likeness to the quality circles now proliferating throughout American industry. However, they differ in some fundamental respects, one of which is their nonvoluntary nature. Unlike quality circles, everyone at Paul Revere, from

President Aubrey K. Reid to the most recently hired clerk, is on a quality team. The quality teams have an average of 10 members, and in 1984, the first year of their existence, were a near mirror of the organizational structure. In virtually all cases, the quality team leader also was the "natural" work unit boss.

The nonvoluntary characteristic is not the only thing that differentiates quality teams from quality circles. Both quality circles and quality teams are charged with the responsibility of identifying areas for improvement. Both also define solutions. The voluntary quality circles, however, must then schedule a presentation with management and offer their findings for management's approval. Quality teams implement their recommendations and then, through a computer program available at 150 terminals throughout the home office, notify a group called "Quality Team Central."

GREATER RECOGNITION

Quality teams also receive company recognition far beyond that suggested by quality circle advocates. The system is tiered, with increasing awards as defined plateaus are reached. The plateaus are:

- Bronze = 10 certified ideas, or $10,000 in annualized savings
- Silver = 25 certified ideas, or $25,000 in annualized savings
- Gold = 50 certified ideas, or $50,000 in annualized savings

Considering that the average quality team in America completes six ideas a year (against four or five in Japan), these plateaus are ambitious.

Of the 127 quality teams at the company last year, 126 made it to the bronze level, 101 went on to the silver, 76 earned gold status, 10 were double-gold, and one team, with 150 ideas implemented and certified, was designated a triple-gold team.

When a team reached bronze status, each member of the team was presented with a bronze pin. Silver brought a silver pin and each member was permitted to choose a gift from a catalog. A more expensive catalog and a gold pin rewarded the gold teams. The double and triple-gold teams earned another item from the gold catalog.

Teams are encouraged to meet for a half hour each week and to concentrate on ideas that will improve the quality of their work—in fact and in perception. "Quality in fact" is defined as doing things to meet your own specifications; "quality in perception" means doing things to meet your customers' expectations. A "customer" is defined as anyone to whom you provide products, service or information.

After a meeting, the quality team leader can enter any new ideas on the computer through the Quality Team Tracking Program, or change the status of any ideas that have been put on hold, deleted, or implemented. This program serves to make the whole process "user-friendly" by taking the place of any paperwork that might otherwise have to be generated. Once an idea is noted as "implemented" on the computer, a Quality Team Central analyst (one of four persons trained as productivity analysts and work measurement specialists) contacts the team leader to arrange a time for certification.

The certification of an idea is necessary for it to count toward earning recognition for the quality team—and it serves as a means of keeping track of all the changes going on within the company. Additionally, the analysts check to ensure that any necessary coordination has taken place, and that no unexpected side-effects will result from a team's idea.

In just short of a year, 7109 ideas were entered on the QTTP. Some of the insurance-particular ones were:

- Dental Claims—Increase authorization limits for medical claim examiners from $3000 to $5000 (savings: $10,144 in time annually).
- Individual Claims—Streamline procedures for requesting independent medical examination on selected claims.
- Investment Department—Develop automated systems to replace manually generated reports for investment mortgages (savings: $10,000 annually).
- Insurance Operations and Purchasing—Ship materials via truck rather than using air for Canadian office (savings: $27,000 annually).
- Legal Department—Review disability reinsurance treaties and procedures to limit unintended exposure to unusual contract and compensation damages and attorney fees (savings: $150,000 on one case).
- Financial Services—Reduce complement through implementation of ideas and reorganization of workload.
- Information Services Department—Delegate work to a lower level person (savings: $9875 in salary annually).
- Individual Claims—Use reusable nylon priority mail bags to send material to field claim representatives in place of manila envelopes (savings: $9200 in postage and envelopes annually.)

Reviewing the full list of ideas invariably produces comments such as, "They should have been doing that all along." This point is not disputed, but it is overridden by a more important point: these things weren't being done, and they are now; and employees are committed to the changes because they suggested them.

The quality teams entered 1985 with a backlog of more than 2000 ideas. About 25% of the team leaders have stepped aside to give others a try at leading, and some teams have been realigned to fit special working relationships. The recognition system has been altered to increase the options available to members of successful teams. Instead of choosing from a gift catalog, employees receive gift certificates redeemable at more than 20 stores and restaurants or tradeable for L.L. Bean gift certificates.

As a point of comparison, consider this. In 1983, the corporate suggestion system, a conventional program with what was considered to be average success, elicited 216 suggestions. Of those, 86 were implemented, with an annual savings of just under $41,000.

The quality team approach defined by the Quality Steering Committee must be credited with the difference in savings since the opportunities for personal reward are lower (the 1983 system rewarded 25% of first-year savings to suggestion givers, with a minimum of $25). Being trusted to take control of their own work, while still being very much responsible for the quality of that work, has inspired employees to take an active part in its steady improvement.

VALUE ANALYSIS

Unlike the homegrown quality team component, the Value Analysis workshops were initially consultant-driven, with employees serving as assistant/apprentice facilitators. After several months, the company was able to assume complete responsibility for the conduct of the Value Analysis workshops.

When the Quality Has Value process was first being defined in mid-1983, a major decision was to proceed with the Value Analysis segment and the quality team effort simultaneously, rather than sequentially. The winning argument was, "Why try to do things right (quality teams) if we're not doing the right things (value analysis)?" Once that decision was made, the consulting firm of Robert E. Nolan was invited to have a representative attend all meetings of the Quality Steering Committee so that we could ensure that the workshops would indeed be complementary to the overall thrust of the process.

The value analysis approach used by the Nolan Company, and now adopted by Paul Revere, is to study one functional department at a time by convening a series of meetings consisting of the decision makers from that department. The value analysis workshops do not directly involve every person in the company, but everyone feels the impact of the decisions coming out of the workshops. The normal workshop requires two three-hour meetings a week, for six to eight weeks.

The first step is to define the primary objective of the department in two words; one verb and one noun. In a paragraph or two, virtually everyone can define his or her purpose for being and make it sound both mysterious and vital. Doing it in two words tends to focus the discussion considerably. People become very careful with their words.

The next step is to name all functions performed by the department and to then determine how, and in what sequence, these functions contribute to the accomplishment of the primary objective. Some functions fall out at this point, especially duplicate or obsolete functions that have endured because they always have been done that way.

At the beginning of the Quality Has Value process, the president of the company made the promise that no one would be removed from the payroll because of the quality process. If a job is eliminated, another is offered within the company.

Evaluation of the functions—both of how important and how well they are done—is followed by brainstorming optional solutions. The brainstorming is done in a restraint-free atmosphere; no personnel or monetary considerations are allowed. A realistic appraisal of the brainstormed ideas leads to a list of recommended changes that may be written by the facilitator but is signed by the workshop participants. Nothing goes into the report with which the participants are not comfortable. It is a list of their recommendations on how to improve the quality of their department.

Some examples of results in 1984 and projected worth to the company when the ideas are fully implemented follow:

- Claims Department—(1) Improve coordination between claims examiners and the legal staff to resolve suits in a more timely fashion and to pursue government benefits on behalf of a claimant ($150,000 plus per annnum). (2) Increase the usage of arbitration versus suits, reducing legal fees ($80,000 per annum).
- Actuarial Department—(1) Create administrative computer systems with imbedded actuarial models and statistical reporting ($100,000 per annum). (2) Standardize the pricing and forecasting methodology used for all product lines to permit better product management decisions ($400,000 plus in increased earnings).
- Pension Services—Contract annual pension plan servicing to an outside service ($80,000 per annum). This suggestion involved dropping nine employees, all of whom were offered equivalent or higher positions within the company.
- Underwriting Department—Restructure the underwriting and issue departments from a production line to a team concept where each team processes all aspects of the underwriting and issue functions for a specific group of field offices ($134,000 per annum).

In some cases, the savings or increased revenue are "hard", i.e., bottom line, dollars. In others, the money is "soft" or "excess-capacity" type savings which will allow the same number of people to handle an increased amount of work due to increased efficiency.

The third aspect of the Quality Has Value process, the surveys, reached everybody during the first year—

employees in the home office and field force, and policy-holders. The general message from the several surveys was that we were in fairly good shape, but there was room for improvement.

One of the uses of these surveys was the establishment of several employee committees to study ways to improve areas within the company identified as being of concern. Another was the establishment of goals for improvement in the perception of policy-holders and other customers. The results of the year's surveys, both the ones generated by the marketing research branch and industry surveys, will be studied closely to see if improvement is being made and to redefine the approach to improvement, if appropriate.

The hoped-for change in corporate culture, as a result of all this activity, has begun, "Quality" is a part of the company's business vocabulary. The remaining cynics are much quieter now as the results continue to mount up.

The company is, quite simply, a better place to work. A director recently pointed out that previously, if you saw three or four clerks gathered around a desk during working hours, you assumed they were "goofing off" and immediately dispersed them to their own desks. Now, however, there is a very good chance they are discussing a way to improve their own efficiency (which may well have been what was going on before, but now it is an expected, accepted thing.)

KEY TO THE PROCESS

This willingness to trust lower-level employees is symptomatic of the culture change. In fact, that trust and the feeling that "we're all in this together" are perhaps the real keys to the whole process.

In recent decades in American industry, all upwardly mobile managers have been taught, implicitly or explicitly, that wisdom comes from on high; that the way to get ahead is to "listen up and proclaim down" the corporate ladder. The Quality Has Value process is teaching people to "listen down." We realize now that a wealth of knowledge about the insurance business—about things big and small that truly affect the bottom line and the customer's perception of the company—is waiting to be tapped at levels far below vice president.

Improved employee morale is a delightful side product of the Quaity Has Value process, but the *raison d'etre* of the program is that it is providing the company with both short-term gains and long-term strength.

Best's Review, *June 1987*

THE POLICY IS STILL QUALITY

Four-and-a-half years after its impressive beginning, the Quality Has Value process at the Paul Revere Insurance Group is not only flourishing, but is well on its way to becoming both a permanent element of the company culture and a model for service industry companies throughout America.

The original fundamentals of the Quality Has Value process have proven themselves to be sound and have remained virtually unchanged. The two most obvious "evolutionary" changes thus far have been the strong involvement of the field force and the switch from a central assessment scheme based primarily on meeting pre-assigned, numerically measured levels of achievement to one based on the self-assessment of progress by the 125 quality teams in the home office.

The bedrock of the process can be found in the definitions of its major terms (such as "quality" and "customer"), its nonvoluntary nature, and the concepts of trust and gratitude. While some short-term programs that have been initiated under the umbrella of the Quality Has Value process may not have been as successful as others, keeping to the basic principles of the process has helped to keep its momentum rolling.

At Paul Revere, "quality" is defined as consisting of two interrelated concepts: "Quality in Fact" and "Quality in Perception." While these two ideas can be discussed independently, it should never be forgotten that a successful effort requires them to work in combination.

"Quality in Fact" is achieved when the specifications of a job—as they are understood by the person or unit performing that task—are met. In other words, when you accomplish just what you set out to do, you can claim "Quality in Fact." The problem comes when you find out that no one else cares. "Quality in Perception" is attained when the customer believes that the service or product being offered will meet his or her expectations. That belief alone will not carry a company far, however, if the product or service—once purchased or received—falls short of expectations.

To make the definitions useful in the day-to-day world, a consistent definition is also needed for "customer." A customer is anyone to whom an individual or a unit provides a product, a service, or information. Anyone. A consumer can be a customer, but so can Mary at the next desk, and/or Gene who's in the next department.

The "trick," then, is to talk with customers, to learn their expectations and set specifications to match—and then to measure up to—those specifications. If someone is incapable of meeting the specifications which will satisfy the customer's expectations, then that customer must be told and some common ground must be sought.

All of these definitions have been used repeatedly during the years since the quality process was officially launched on January 13, 1984.

INVOLVING EVERYONE

The quality process at Paul Revere involves every person who draws a paycheck from the company, and this has helped immeasurably in setting the process apart from others similar to it. The fact that enrollment on a quality team is non-voluntary is rooted in the answer to a simple question: "Whom can a company afford to exclude from an effort to improve?" One observer of the Paul Revere process noted that if an organization can identify someone whose potential improvement is of no value, the best place to begin may be with firing that person, regardless of his or her position.

Actions, of course, do speak louder than words. The president of Paul Revere is on a quality team—and so is the newest hire. At new employee orientation classes, the newest members of Paul Revere are often greeted with the words, "Congratulations, you're on a quality team!" The trust is also more than mere words included in a recruiting brochure. It is extended in such a way as to be a call to action. Quality teams are automatically granted authority commensurate with their responsibility.

In other words, if a quality team wants to change something that falls within its area of responsibility, it should assume the authority to make the change. It should not wait to make presentations to management, or to ask for any permission. No delays should come into play at all.

In 1984 and 1985, the main thrust of Paul Revere's efforts to say "thank you" to participants was the neatly structured "Olympic medal" approach. Reflecting the fact that the first year of the Quality Has Value process was also an Olympic year, the three levels of recognition were named Bronze, Silver and Gold. To qualify for Bronze, a quality team (average size: 10 members) had to implement 10 ideas or a lesser number of ideas that had an annual worth (hard and soft dollars combined) of $10,000. Silver required 25 ideas or a worth of $25,000; Gold, 50 or $50,000.

A few months after the process began, the director of Quality Team Central discovered that the average quality circle in America completes only about six ideas a year. But it was too late to alter Paul Revere's standards accordingly. Furthermore, four teams had already reached the Gold level before May, suggesting that this was indeed a definitive new step in the effort to involve employees in the improvement of an organization.

In 1986 one home office quality team reached the Septuple Gold level (i.e., fulfilled the requirements for "Gold" seven times in 12 months). In the home office in 1985, the 1,250 employees implemented 5,702 separate ideas. Putting these findings in the proper perspective, the "implemented ideas per employee" at Paul Revere that year was *350 times* the national average reported by the National Association of Suggestion Systems for implemented ideas per eligible employee at companies with suggestion systems.

Although the process was thus doing very well in terms of producing a wealth of ideas, the participants and leaders continued to try and improve upon it. At a meeting of the quality team leaders in November 1985, the participants agreed that the teams should go to work on designing more long-term improvements. This approach would help to free everyone's creativity from the previous, more structured system (which relied heavily on exact counts of ideas and money saved.) Participants would now become more independent, setting their own goals and beginning to assess their own progress along the way.

This was the beginning of self-assessment by Paul Revere's quality teams. At the beginning of the year, each quality team was required to define its quality goals for the upcoming 12 months. These goals were to be as easy to measure as possible and to be ideas that the teams could either accomplish themselves or could implement with the cooperation of another team that shared in the goal.

Every three months, a quality team would receive a self-assessment form. After the indentification information on the top of the sheet came two questions: "How do you rate yourself?" and "Why?" More specifically, the team was asked to circle the level at which the members felt they had performed—with relation to their quality goals—during the previous three months. The choices were Pass, Bronze, Silver and Gold. They were then asked to describe the actions that they had taken to justify their rating.

The self-assessment sheets were reviewed by the same people—Quality Team Central—who reviewed and certified every single implemented idea. In an average month (one-third of the teams received self-assessment sheets each month), two assessments would be downgraded and one would be upgraded. Downgrading occurred only after a long talk with, and the concurrence of, the team leader.

THE PROCESS GROWS

Giving even more autonomy to the home office quality teams doubtless contributed to the electrifying growth in participation in the process by the field force in 1986. "Pym's Rules" also help to explain why the field went from 1,002 implemented ideas in 1985 to more than 4,000 in 1986.

According to Jeff Pym, director of the quality effort in Paul Revere's Canadian Head Office, every individual and unit will go through three phases during the growth of a quality process. These are: (1) What can you do for me?, (2) What can I do for me?, and (3) What can I do for you?

The field force, including sales people and administrative/clerical support personnel, was formally enrolled in the quality process in mid-to-late 1984. Each field office received an initial indoctrination on the process by a regional sales vice president as he made his way around his various offices. Ideas came slowly as the field force, veterans of many highly touted, short-lived programs, waited this one out. The system–and, specifically, the people in the home office–were going to have to prove that the process was worth the investment of any time. In short, they wanted to know what could, and would, be done for them.

IDEAS FROM THE FIELD

Quality ideas from the field came in two categories. The ideas could either be ones that were implemented in the specific office in order to improve its own quality, or requests for action in the home office. The latter could either apply solely to the submitting field quality team or have nationwide implication.

As 1985 evolved, field quality teams gradually became convinced that this quality process was truly here to stay, that the people in the home office really were on their side, and that specific things were being done on their behalf. And, when a team's ideas were turned down, Quality Team Central (the conduit for all ideas from the field) insisted that the declining person in the home office give a detailed explanation for the refusal.

By 1986, the majority of the field teams were, for the most part, operating at the second level of Pym's rules, while the home office quality teams were asking with increasing frequency, "What can we do to help?" The result was that, in 1986, 4,100 field quality ideas were implemented in addition to the 4,000 home office quality ideas.

The most striking single example of the change in the relationship between the home office and the field was the introduction of a completely new policy proposal system in late 1986 and early 1987. The new system was based on 117 field quality ideas; the system was constructed with the field force, rather than simply for them.

A second example can be found in a quality team called the "home field advantage." This team consists of several directors and vice presidents from the home office, plus field representatives from each of Paul Revere's three major distribution channels (direct, brokerage and group). Their self-appointed charter is to improve communications between the home office and the field. More than one meeting begins with a field representative saying, "Let me tell you what you did to us this month..." Often, the problems are resolved before the meeting is over.

CELEBRATING SUCCESS

Characteristic of the Paul Revere Quality Has Value process from its beginning has been its insistence on the importance of recognition, gratitude and celebration. In the years since the process was first initiated, several variations of showing appreciation to participants have been tried.

One of the best received shows of gratitude was the "thank you for a cactus" program of February 1986. The rules were simple. Anyone who received a "thank you" letter from any of his customers (remember, a customer need not be a consumer) could bring a copy of the letter to the office of the Director-Quality Team Central and exchange it for a cactus...with a sticker on its pot proclaiming "I'm stuck on quality." Was this undeniably corny? Yes. But it was also a great deal of fun. And it resulted in reminders of the quality process all over the building–reminders that would live almost until a baseball bat was taken to them.

Perhaps the most transferable lesson that has been learned about quality processes is that they are neither easy nor cheap, but they are well worth the effort. In the first three years of the Quality Has Value process, the premium income from the company's primary product increased by more than 95% while the staffing level rose by 4%.

A quality process needs a well-developed plan to begin with and a willingness to change the peripheral aspects of that plan while adhering to its basic principles. It will need a "chief mechanic" who, with his or her staff, is responsible solely for the success and continual evolution of the process.

Paul Revere is not yet finished. As stated in *Commit to Quality*. "The decision to begin a quality process is a decision to allow a revolution." As with all revolutions, this one takes time. One basic rule of the "quality game" is worth remembring: the first company to get it figured out exactly right gets to keep all the money.

Best's Review, *June 1988.*

QUALITY INVOLVES EVERYONE: HOW PAUL REVERE DISCOVERED "QUALITY HAS VALUE"

With all the hype about service quality currently circulating–and all the theories that defy common sense being peddled by an astounding number of "TQM consultants"–perhaps it is best to begin with a few simple truths.

- The primary reason for implementing and maintaining a quality process in a non-government organization is the making of more money. Done correctly, the effort will also increase customer satisfaction and loyalty and employee morale, but the bottom line is the bottom line.
- Service to external customers will, as a rule, never exceed service to internal customers, particularly over any sustained period of time.
- No one employee can ensure the reputation of an organization, but any one employee can ruin it. Thus, it is illogical not to involve every person on the payroll in a quality process.

This article will use as its continuing example the "quality-has-value" program that was implemented in the Paul Revere Insurance Group in 1984 and that, in an evolutionary form, continues today. In 1983, Paul Revere's market share in the USA for its primary product, non-cancellable individual disability income insurance, was 11.8 percent, good enough for second place in a wildly competitive market. By 1994, market share had risen to 18.4 percent, with no competitor above 10 percent. Formal customer service surveys were at a minimum prior to the initiation of the quality process but surveys taken once the process began have shown a steady increase while, at the same time, the number of complaints (to include the most formal sort, i.e., lawsuits) has decreased dramatically.

It should also be admitted at the outset that much of what is about to be said (and has been said in numerous speeches and articles in the USA, as well as in the book, *Commit to Quality*) should most fairly be labelled "retrospective theory." Much of what was attempted at Paul Revere had no precedent or underlying theoretical base. It seemed logical, it made sense, and the employees understood it, so it was done. After it had worked out so well, the question was asked: "Now, why did that work so nicely?" Result: retrospective theory.

This idea of retrospective theory adds weight to the notion that effective leadership requires time to reflect. Continuous activism can lead too easily to continuous turmoil. Once a new direction is understood, agreed, and has proven to be beneficial, building on success becomes the logical path. To do that, the past needs to be understood.

An example of retrospective theory would be "Pym's phases of quality" and "McConville's corollary." After the first two years of the Quality Has Value process, Jeff Pym (he directed the quality process in Paul Revere's Canadian headquarters) pointed out that people and teams both appeared consistently to go through the same sequential phases as they grew more comfortable with the quality process. These phases were:

- What can you do for me?
- What can I do for me?
- What can I do for you?

About two years later, Joe McConville (the then-new director of the process in the USA) added a corollary, a fourth sequential step, which had become evident during the intervening years: What can we do together?

Some people–and teams–moved through the four phases in a matter of days or weeks; others took many months. As a rule, for instance, sales agents in offices detached from the headquarters offices spent a long time in the first phase before trusting that this quality process was not just another sales promotion that would die at the first crisis.

Having the behavior codified by Jeff Pym and Joe McConville helped executives to understand what had happened and it helped in the planning of future developments.

The original decision by the leadership of Paul Revere to "try quality" was for the purest (for these purposes) of reasons: to make more money. While still quite profitable in early 1983, it had fallen to number two in its chosen field, and it did not like it. On top of straight business motivation, its pride had been stung; it had, after all, been the industry leader for decades prior to 1983.

The fall from the lead was attributed in large part to having let attention wander, from assuming that the main product would take care of itself. At one point in preceding years, Paul Revere had even owned a steel mill. The company that had supplanted Paul Revere at the top, incidentally, was the Provident—which has now returned to the number two position, with less than a 10 percent market share.

Under co-chairmen Chuck Soule (the senior vice-president for operations, who would succeed to the presidency in 1990) and Bill Pearson (vice-president for human resources), the Quality Steering Committee began meeting in May 1983. The president of the company, Aubrey Reid, was not on the committee but he did promise to act on their recommendations, a promise he kept.

At the outset of the definition of a quality process, the question that is most frequently asked (in some form) is: "Who should we involve?" It is a question that leads to all manner of lengthy, defendable, wonderfully precise discussions with the end result usually being about 10–20 percent of the workforce involved "for now," with the rest to be included at some undefined future date.

At Paul Revere, the question that was asked was:"Who can we afford to exclude?" It was quickly agreed that the answer was "nobody," and the meeting continued. In fact, the decision to go immediately to 100 percent involvement was not even noted in the minutes of the meeting. It was, after all, so logical that it did not seem like a big deal. The only remaining question was exactly how to do it. When asked months later about the thinking of the committee at the time, Ken Hedenburg, Chuck Solue's assistant, replied: "I think we thought we were doing quality circles."

The one point over which there was extended debate in the quality steering committee was whether to begin with the value analysis (to put this term in current perspective, reengineering can be thought of as value analysis with an attitude) effort or the quality teams (the vehicle chosen for employee involvement) effort. After several relatively heated discussions, Chuck Soule and Bill Pearson sat down one evening to come to a conclusion.

The point was that value analysis tended to answer the question: "Are we doing the right things?" while the quality teams could be expected (and directed) to address the question: "Are we doing things right?" After agreeing that it would be of only minimal gain to do the right things if they were to be done poorly and, on the other hand, it would be of even less benefit to do things well if they were the wrong things to do, the decision was to do both right from the start. Once again, logic prevailed.

The value analysis effort would consist of a series of workshops in which the management teams of each of the major divisions and departments of the company would be led through an analysis of the structure of their organization—to see if it was built correctly to best handle the workload, problems and challenges currently facing it. Since any one workshop would take 6–8 weeks and they could conduct only three or four at a time, it was estimated that it would take at least two years to get completely through the company.

Everyone was to be on a quality team—from the president of the company to the person hired yesterday in the mailroom. Initially organized by "normal" work groups, it was anticipated that membership would begin to shift fairly quickly as the need for cross-functional teams became obvious. There were no absolute rules concerning team make-up. The only absolute rule was that everyone had to be on one.

Another benefit of not having outside help was that the Paul Revere Quality Steering Committee did not know that it was supposed to take years to define, initiate, and benefit from a quality process. At the time, there were 1,250 employees located in Worcester, Massachusetts, 1,250 located in sales offices throughout the USA and another 500 employees in Canada–split between their headquarters building and sales offices throughout the country.

The committee's first meeting was in May 1983. On 13 January 1984 the process was launched in the USA and Canada. In fact, the value analysis workshops had already begun in Canada; 13 January marked the official beginning and the actual initiation of the quality team effort in both countries.

During those eight months, the process had been defined to include the quality team mechanics, a communication system, a training curriculum, and a recognition scheme. Quality team mechanics included the training of the quality team leaders and the establishment of a tracking system and support staff.

The staff who ran the process in the USA consisted of a director (the author of this article), a part-time secretary, and four quality analysts (former productivity analysts). In Canada, directing the process was a part-time job for Jeff Pym, and he had one assistant.

How good could a process be that came together so quickly? The first recording of a bottom line benefit took place 90 minutes after the official launching of the process. It was admittedly a small annual saving, a matter of a few hundred dollars saved by a team of clerical workers through the elimination of a wasteful and bothersome procedure, but the speed of the action was indicative of the ready acceptance of the process.

The initial monetary investment in the process–the cost of the team leader training programme, printing of materials, wages paid to the consultants who were helping to establish the value analysis workshops (and training the Paul Revere employees to take over)–was completely paid back within six months.

From that point on, costs were always exceeded by savings and added income by an order of magnitude. Costs included the programme of recognition, gratitude, and celebration; the development and introduction of new training courses; and administrative costs–to include the salaries of the people whose sole job was to administer the process.

After the first three years of the quality-has-value process, the income to the company from the primary product had increased by 95 percent while the number of employees on the payroll had increased by only 4 percent. After four years, Paul Revere was one of only two finalists in the Service category for the Malcolm Baldrige National Quality Award in that award's inaugural year.

The failure to win the Baldrige can be attributed at least in part to the fact that the Quality Has Value process had deliberately avoided installing any detailed, manufacturing-like measurement systems. This decision had been due to the fact that it had been only a few years since there had been a rather vigorous productivity analysis programme imposed at Paul Revere which had served mostly to make everybody mad. Quality teams were adding more and more measurement systems with each passing year, but they fell short of the Baldrige expectations in 1988.

After receiving the feedback report from the Baldrige examiners, Paul Revere incorporated—or scheduled—all changes and improvements which it deemed appropriate. The decision not to re-apply was very much in keeping with the spirit and intent of the Baldrige. The primary object is not to win; it is to learn and improve. Former examiners tell of reviewing applications which left questions unanswered (or answered by something like: "We haven't addressed this point yet"). The companies were obviously applying only to get the objective assessment and the feedback report, which then serves as the blueprint for their next round of internal improvements–whether or not they ever re-apply.

To explain the quick–and sustained–success, both the mechanics of the process and the inter-personal aspects of the organization need to be addressed.

The quality team system that was set up was simple to understand, transferred power to the appropriate levels, and had the obvious involvement of every person in the organization. Teams were told that they were given 30 minutes per week—on company time—to meet to discuss how to improve whatever it was they did. Most teams did meet for a half-hour each week, regular as clockwork. A small number of teams met for 60 minutes every other week. One team that included members from several locations met for two hours once a month. Others held a near-continual series of small meetings, often in the hallways. The point was that people began talking to one another about how to get better. Meetings were not monitored or questioned. Results were, but, even then, it was primarily on a "What can be done to help you have better results?" basis.

COMMON VOCABULARY

Making the conversation possible was the use of a common vocabulary:

- "Customer" was defined as anyone to whom service,

product, or information was provided.

- "Quality" was defined as having two components: quality in fact and quality in perception. The first was fulfilled by doing exactly what you set out to do, what you had promised – to yourself if no one else. In other words, to achieve quality in fact meant to meet your own specifications. The second, quality in perception, was fulfilled only when someone else believed that the service, product or information being offered was exactly what they wanted, that it met their expectations. To claim to have, or to be, "quality" required both.
- A "quality process," then, was agreed to be a continual, proactive effort to determine the customer's expectations and compare those to your specifications – and determine if there was a match. If there was no match, then it was the responsibility of the provider to do something. The first choice was, of course, to change the specifications, what it took to achieve quality in fact... more commonly known as "the customer is always right."

Sometimes, however, the customer is under-informed. Sometimes it is not currently possible to meet the customer's expectations. In that case, the provider's option is to initiate a discussion–and to educate the customer's expectations. By shifting the expectations (and, most probably, the specifications as well), common ground can almost always be found–a point at which the specifications and expectations meet. Once there, the provider's job is straightforward.

The next step is less obvious: ask again what the expectations are. It is this step that makes service quality a more difficult challenge than manufacturing quality, since customer expectations tend to change more rapidly and more frequently in the world of service than in the world of manufacturing.

This vocabulary gave everyone a non-confrontational way to assess problem areas. If one person can ask another "Please tell me again what your expectations are so we can compare them to my specifications and we can figure out what needs improvement," the chances of beneficial change are far higher than if they ask "Now what's wrong with you?" With this vocabulary, the problem becomes an impersonal thing, not a character flaw in one or the other person.

Quality teams at Paul Revere were empowered before that word began its meteoric rise to (and short stay at) the top of the cliche list. The definition used was short and sweet (and borrowed from the military): authority equal to responsibility. No more but no less. In commoner terms, if a team's collective tail was on the line for something, members automatically had the power to effect a change.

By giving teams real power, the process focused the teams' efforts on the problems and opportunities immediately at hand. Hundreds of quality circle efforts in the USA got bogged down by circles suggesting all manner of changes in matters about which their knowledge was, at best, superficial. That is what happens when there is no power to act. If you cannot really do anything about it anyway, why not have ideas about how someone else should do their job?

The belief that external service rarely, if ever, exceeds internal service and that every interaction with the "ultimate" customer, the policy-buyer, is at the end of a chain of provider-customer mini-processes involving internal employees gave strength to the idea of engaging everyone in the process and giving them access to power that enabled speedy improvements.

As the process got under way, one of the many shared insights was that, for the majority of the people in the headquarters office, the primary customer was not the individual out there who actually purchased an insurance policy, it was a person in the sales office–alternately administrative staff members and sales people.

As a result of this realization, one of the offices in the headquarters decided to begin a series of "How are things going?" calls to the sales office. It was recognized that this decision (to initiate phone calls rather than waiting for the phone to ring) would increase their phone bill but they felt that it was the right thing to do for their customer.

The result? Happier customers and *lower* phone bills. It turned out that by getting early warning of brewing problems, the headquarters personnel were able to avoid many problems, thus eliminating a series of long, frantic, unhappy phone calls a month or two down the line.

Another example of the new interplay between sales offices and headquarters was the monthly sales office survey initiated in about the fourth year of the process. The two primary questions were "In the last month, what process at the home office served you badly?" and "In the last month, what person or people served you well?" By keeping the problems impersonal and the gratitude very personal, it was possible not only to concentrate resources on the problems that were causing the most concern but also to ensure that deserving individuals were recognized and thanked.

That this increase in co-operation and mutual concern between all segments of the company did reach out to the buying public is evident in the steadily increasing sales figures. Besides the increase in market share noted above, in 1993, Paul Revere had its fifth consecutive year of record profits. In addition, while the vast majority of insurance companies in the USA have reduced personnel at some point in the past ten years (through downsizing, rightsizing, correctsizing, or by simply firing people), Paul Revere has grown steadily.

LEADERSHIP, PARTICIPATION, MEASUREMENT

By balancing these three major components of a quality process and by maintaining control of the mechanics and direction of the process at all times, Paul Revere blazed a trail that surprisingly few have followed.

The lack of imitators is no doubt due to three factors:

(1) The process had been very much an in-house invention, so there was no large consultant company trumpeting the news and offering to do likewise for any other lucky/discerning client.

(2) After brief consideration, Paul Revere decided not to establish its own consulting firm. Although that was a path chosen by several other firms in the USA, Paul Revere (perhaps remembering why it had slipped to number two initially) decided to stay focused on its main strengths.

(3) Paul Revere made no particular effort to advertise its success, content with increased business, happier (and more plentiful) customers, and more satisfied employees. As one measure of the latter, by 1988, Paul Revere's employee turnover rate was half that of the industry average. It would have been lower but there was no standard method for calculating the impact on turnover of employees who left and, having discovered that their new employer did not want to hear their ideas, came back.

It is the unfortunate truth that top management teams of most US service companies have chosen to follow the siren call of large consulting firms and invest money instead of themselves, involve only a small percentage of their people, and order up enough measurements to keep dozens of staffers busy getting ready for the next presentation.

The first step in the initiation of a complete quality

process is to believe that Personnel has been hiring adults and that brains are evenly distributed within an organization, i.e., one per person. Next, the senior management team must believe in itself, its knowledge of its own business and its ability to adapt universal TQM principles to its situation.

The prize is obvious. The first organization that figures it out in each industry gets to keep most of the money. In order to stay ahead of the pack, the senior leadership of Paul Revere has guided the quality-has-value process through a number of evolutionary steps. The underlying theory is that a process that urges continual change must itself be open to change or the inherent hypocrisy will doom the process.

For example, an effort that may well have first been seen as process-driven has become more market-driven. Beginning as it did was correct because it was necessary first to ensure that all employees understood and embraced the process and that all were well served by it. (Remember one of the basic truths: external service rarely exceeds internal service.) That the evolution would take several years was understood and accepted for two reasons:

(1) bottom-line benefits from the beginning;

(2) since integration of "quality" into company practices was to be "from now on" there was time to do it right.

The all-important external customer (it is their money, after all) was served indirectly from the beginning since the company was intent on improving its internal practices, allowing claims to be processed more quickly and accurately, enabling agents to deliver policies more quickly, and ensuring that agents who dealt directly with policy buyers were better supported. That improvement in service to customers is a continuing result of the process.

The quality-has-value process today looks far different from the process launched on Friday, 13 January, 1984. But the principles remain: leadership, participation, measurement. And the success continues.

Managing Service Quality, Vol. 5,*Nov. 2*, p. 19–24.

11

INDUSTRY-SPECIFIC EXAMPLES

Adopt the principles; adapt the practices. When looking at another organization's quality process—either through reading about it, taking a quick tour, or being involved in a formal benchmarking or partnering relationship—keep those two guidelines in mind. No one company quality process will perfectly fit any other company, but everyone can learn from virtually everyone else. One of the most valuable characteristics of the worldwide quality revolution has been the amazing openness with which organizations have treated their quality efforts. There are few secrets kept. Unfortunately, not everyone takes advantage of the knowledge available.

This chapter includes examples and ideas from several types of organizations: health care, software, communications, computers, and the military. In some cases, the lessons offered are positive ones; in others, the article's value is to point out errors to be avoided.

Escalating costs, increased public attention, and population growth were some of the catalysts for the surge in interest in quality within the health care industry—and great strides have been made. "Will Continuous Improvement Work Here?" offers a sure answer with regards to health care: Yes.

"Breaking New Ground" addresses quality in the communications and computer industries; "The Right Question" looks at the world of software. Again, the principles carry over although techniques change—just as they do when comparing service organizations to manufacturing concerns, as is discussed in "Total Service Quality."

The US military has made several attempts to take advantage of quality principles to help it prepare for a future in which the only certainty appears to be declining resources. Unfortunately, efforts thus far have been flawed, as will become evident in "Total Quality Leadership or Partial Quality Management." The article reviews the seven elements of a Complete Quality Process and examines its implementation in the US Marine Corps. The results are not encouraging.

WILL CONTINUOUS IMPROVEMENT WORK HERE?

Since the beginning of this nation's history, Americans' experience with healthcare has been characterized by increased capabilities and accessibility. Good healthcare for the majority of Americans, once a dream, became first a probability and then a reality. In recent years, however, the upward spiral seems to be flattening. Increased capabilities (often the result of expensive medical procedures) and increased accessibility (financed for the most part by employer-provided or government-provided insurance plans) are both threatened by economic considerations.

Everyone feels the crunch. Users are being asked to pay part of the cost of healthcare plans; insurers are turning to cost/benefit analysis to determine what will or will not be covered; medical professionals are faced with ethical dilemmas unheard of a generation ago. While the world of limited resources has not gotten to the "either one-heart-transplant or 50,000-polio-vaccinations" worse-case scenario quite yet, decisions, very tough either/or decisions, are being made. And nobody's happy about it.

The need to find a balance between controlling costs and providing optimal medical care is urgent. How many "million–dollar" babies, how many heroic procedures, how much catastrophic illness can the medical system support? Society as a group expects all of these needs to be met, but fiscal realities are grim and getting grimmer. Howard University Hospital lost $28 million in 1989 alone caring for elderly patients who were unable to find room in nursing homes. As the baby-boomers age, these pressures will increase.

Alternative healthcare venues are developing—palliative care at home for the terminally ill is one example – as other resources are shrinking. Depending on which authoritative-sounding source one chooses to believe, somewhere between 10 and 40 percent of all hospitals in America are said to be in danger of closing before the end of the century. Competition between hospitals for patients is reflected in advertising budgets unheard of a decade ago. Over $1 billion is spent annually to attract patients, thus cluing patients that they can participate in healthcare decisions and ask for a particular hospital. Out of such statistics are new approaches born.

Healthcare is not the first segment of the economy to be challenged by limited resources, customers with high expectations and long memories about what something used to cost, and government pressure. During the 1980s manufacturing and non-medical service organizations which faced similar problems turned to the concepts of quality and employee involvement as the vehicles for solution to these myriad problems. The quality improvement task force of the Joint Commission on the Accreditation of Healthcare Organizations (JCAHO) is taking the same approach in a draft for its accreditation rules. In the future, it has been proposed that a quality process be required for accreditation.

THE QUALITY IN HEALTHCARE FIT

The core concepts of quality—leadership, participation, measurement, and customer service—will translate to a healthcare setting. As with many organizations, quality does not represent startlingly new practices and thoughts so much as it requires giving a well-defined, and well-supported, focus to ongoing efforts. A brief look at each of the four areas mentioned above reveals that healthcare has some natural advantages in three of them.

Built-in advantages It has become axiomatic that quality begins with leadership. The JCAHO quality task force suggests that: "The organization's leaders set expectations, develop plans, and implement procedures to assess and improve the quality of the organization's governance, management, clinical, and support services." The advantage a healthcare provider enjoys in setting quality expectations is immediately obvious. Its employees are already focused on the ultimate goal of any quality process—customer satisfaction. From those in charge of keeping the patient's room clean to the nurses to the admitting staff to the physicians, the rallying point has always been the good of the patient.

There is, however, a concomitant disadvantage. While many businesses have had to shift gears from a focus on the bottom line to far more idealistic values based on cooperation and service, healthcare providers will have to shift gears to an acceptance that the bottom line is not unimportant. And to understand that cost control is not a dirty phrase or a necessary evil.

Clear definitions needed A clear vocabulary can help. Anyone familiar with Townsend and Gebhardt is aware we very tediously insist that quality is composed of two pieces; quality in fact and quality in perception. The first characteristic is fulfilled when the provider of service meets his or her understood specifications, i.e., he or she does exactly what is understood to be the task at hand. The second, quality in perception, can only be claimed when the customer believes that his or her expectations are being met. In order to know when a customer's expectations are met, the basic vocabulary also includes a definition of customer: anyone to whom you provide service, product, or information.

Most simply put, the object of a quality effort is for the provider to determine the customer's expectations, compare those expectations to the corporate specifications and determine if there is a match. If there isn't, something needs to be done—and quickly. This can mean either raising the specifications to the level of the expectations or, if that is not possible, educating customers so that their expectations move to a more realistic level. Once a match between specifications and expectations has been assured, plans are made to meet specifications consistently.

Contrast these definitions developed in a business setting to the definitions developed independently by Dr. Brent C. James, M.D., for the Hospital Research and Educational Trust and published in an monograph titled *Quality Management for Health Care Delivery;*

> "A 'healthcare delivery system' is a series of interlinked processes, each of which results in one or more outputs. 'Quality' represents an individual's subjective evaluation of an output and the personal interactions that take place as the output is delivered to the individual. It is rooted in the individual's expectations, which depend upon the individual's past experiences and present needs. Quality evaluations therefore arise from, and are part of, an individual's value system. As a value system, quality expectations can be measured and changed over time through educations. They cannot be dictated.
>
> Quality has two main components–content and delivery. Content quality is concerned with the medical outcome that is achieved. Although patients and payers are playing an increasingly active role in evaluating medical content quality, it has traditionally been the province of physicians and other healthcare professionals. Delivery quality reflects an individual customer's interaction with the healthcare system for the patient. Was the hospital clean? Were the nurses caring and informative? Were services delivered rapidly, cheerfully, and with understanding of the patient's individual needs and preferences?
>
> A 'customer' is any individual who makes a quality judgement regarding any output or sub-output produced by a healthcare process, or the personal transaction in which the output was delivered."

Multiple customers... This later definition means, of course, that patients are not the only customers; doctors can be the customers of nurses and/or staff and, depending on the moment and the issue at hand, the nurses and staff may be the customers of each other or the doctors. Communication between customers and providers is paramount, if expectations and requirements are to be brought into line.

Washoe Health Systems Facilities of Reno, Nevada, tells of one team that spent four months defining the word "clean" in order to meet the needs of a variety of customers. The team recognized that what was sufficiently clean in the waiting room would not measure up in the examination room and what was good enough in the examination room did not fulfill the needs of an operation room.

Reducing waste and improving productivity... While quality begins with setting expectations, it continues with plans and procedures that produce these results. As in other businesses, the one great benefit that quality can bring to a hospital or any other healthcare institution is in the reduction of waste.

Another indication that business quality is congruent with healthcare quality is Dr. James' analysis of the relationship between quality and cost. In his monograph, noted above, he relates quality to cost in the areas of waste, low productivity, cost–benefit analysis, new technology, and preventive medicine or environmental healthcare; and he maintains that, "The first two areas—quality waste and productivity—offer potential cost savings while quality is maintained or even improved." These are the same savings discussed in business in the context of cost of nonconformance (or cost of quality).

If a factory operation wastes a certain number of pounds of metal or the time of a percentage of its work force, the results are often indirect. The metal is recycled, a few more folks are hired, the price is raised a bit to cover all this, and life goes on without disturbing the consumer unduly. Hospital waste is not as benign. Equipment, personnel, or medicine, quite literally, may not be available when it is needed to save a life. Resources tied up doing, or redoing, work that should have been completed, transmutes into suffering. This is just one instance in which the hard-nosed business aspects of a quality process are in total support of the altruistic goals of any individual healthcare professional or organizational provider.

Who can you afford to leave out of the quality process? A goal seldom realized in non–health related institutions is 100 percent participation in quality improvement. While it is within the reach of every organization, healthcare providers also have an advantage in respect to participation.

Doctors, nurses, technicians, administrators, and many of the support personnel are a self-selected population; they see themselves as having vocations rather than jobs. Many grew up wanting to be in a healing profession—and they persist, despite long hours, emotional stress, and such vagaries as malpractice suits. This evokes an unusually keen awareness of each individual's contribution—an awareness which can be channeled into quality efforts.

It's the 'who' of quality, not just the 'what'... Involving everyone in quality is mandatory in a healthcare setting; every employee is seen as part of the same experience by a patient and all contribute to the length of time of recovery and the total bill.

The need for all members of the staff to cooperate in sending the patients home happy was underscored by a survey published in *USA Today* on October 11, 1989. When asked why they chose a particular healthcare provider, 50 percent of those polled said their decision was based on the recommendations of friends and neighbors. Only 17 percent based their decision on the more rational "technical reputation of the medical staff." That weighing of factors may, in fact, be a left-handed compliment to America's medical professionals.

The whole institution delivers services to its patient/customers... Patients come to hospitals expecting, for the most part, to leave healthy. They expect the place to be clean. They also expect the food to be bad and fear a bill that's difficult if not impossible to read. If, a week later, patients leave healthy and if the place was generally clean... well, of course; if in addition the food was good... they'll recommend the hospital to all of their friends. If beyond that, patients get a bill they can read... they'll be recruiting new patients for the hospital. Is such an assessment fair? No, but it is what happens.

PROCESS FLEXIBILITY

To say that 100 percent involvement is imperative, however, is not the same thing as saying everyone on the payroll must use the exact same quality process mechanics. Clinical and non-clinical personnel are conscious of the difference in their responsibilities ("You make 'em healthy and we'll make 'em happy"), and such differences call for a wide variety of techniques.

Teamwork and measurement Healthcare providers already use some techniques associated with successful quality processes. For instance, teamwork is a recognized means of achieving quality goals, and clinical personnel are in the habit of meeting in teams to assess recent work.

A quality process, however, would bring a different focus to clinical discussions. Rather than determining which individual fell short, as in the past, a quality process would strive to provide data on a specific procedure and subsequent results (i.e., how long on operating table, tissue removed, post-operative infection, relapse) to help doctors evaluate how to perform such procedures in the future.

The wide use of measurement techniques to track results has long been addressed (and checked to a degree by after-the-fact committees) on the healing side of healthcare organizations, and is a natural advantage in the implementation of organization-wide quality measurements. Measurement can and should become a priority in healthcare for much the same reason as in other businesses. Benchmarking, SPC, Pareto analysis, surveys are all tools that are invaluable when applied consistently and thoughtfully.

Caution: quality measurement is not just a bigger hammer... A cautionary note may be due at this point: concentrating solely on measurement is unacceptable. Healthcare systems that take the new JCAHO guidance as merely a call for a larger, tougher police force with stricter quality control and assurance rules will be missing the point.

Consider, for example, one of the most vexing problems in healthcare—the nursing shortage. A sampling of nurses indicates that they leave nursing for a variety of reasons: lack of respect by doctors, tension with peers, patient overload—all of which produce undue stress. A lot of government-mandated or institution-mandated quality control forms are not going to ease the stress load. Conversely, quality emphasizes communication, teamwork, and trust—any one of which has the potential to reduce the amount of undue stress in the work environment.

Other quality process techniques can mirror regular service industry quality efforts (assuming regular means formal enrollment of every person in the process of continuous improvement). Training and/or recognition, gratitude, and celebration serve the same purposes of reaching all corners of the organization.

Customer satisfaction means no surprises The emotional stakes in healthcare are extremely high in the area of customer satisfaction. This is true whether the issues relate to quality in fact or to quality in perception—and whether the customer is internal or external. Surprises, no matter how small, are not welcome in a hospital.

When the dietitian comes to visit a patient and finds the room empty; when a patient awakens from an operation and feels worse than expected; when a nurse cannot find gauze in the supply room, it takes on an importance not associated with slow service in a department store or cold food in a restaurant.

Patient education... The need to educate customers is especially critical in the case of patients and their families. Many an extended hospital stay and many a malpractice suit are the results of a mis-communication between a doctor and a patient or between the patient and other hospital personnel.

A survey conducted in 1987 suggests that doctors have a long way to go in this regard; only 61 percent of the respondents who had recently seen a doctor were completely satisfied. This dissatisfaction impacts clinical results; a patient who has little trust in a doctor is likely to inhibit the healing process, whether because of stress or a misinterpretation—or even willful disregard—of doctor's instructions.

A quality process is the only option Quality in healthcare has never been optional. If the JCAHO quality improvement task force has its way, so will a true quality process. As the real world of budgets and limited resources and competition crowds more and more into the sanctuary once afforded to healthcare, the institutions which wish to see the 21st century as viable corporate entities must begin now to study the options and possibilities. Each must determine how they will adopt quality principles and adapt quality process to enable 100 percent of the people in its payroll to take part.

Journal for Quality and Participation, *Jan.–Feb. 1991.*

BREAKING NEW GROUND

From deli and donut shops taking lunch orders by fax, to control center computers determining and transmitting course corrections for satellites millions of miles away, to the simultaneous printing of identical daily newspapers throughout the country, the communications and computer industries surround us like no others.

"Hungary, Poland, and the Soviet Union, among others, may spend $100 billion on major efforts to modernize their phone systems over the next decade." – *Business Week*, November 20, 1989.

The universal impact of these mutually dependent industries makes their quality a major concern for all of us. Technical quality assurance procedures, do-it-right-the-first-time philosophies practiced assiduously by planners and designers, quality processes involving all employees and reaching out to customers to more clearly determine expectations and dreams, all contribute: no one approach can be either excluded or solely depended on.

The complexity of establishing quality practices for both industries is increased by their abstract natures. Unlike the railroad lines of yesteryear, communication lines, physical and electronic, transfer their cargo without a physical trail. No train whistle sounds at the crossings. Nevertheless, there are parallels between the transportation lifelines of yesterday and the communication lifelines of today. Just as with the conveyance of goods, the conveyance of information is not only big business in itself, it is also vital to the conduct of business in general. And if the phone lines are today's equivalent of railroad tracks, surely computers are the current surrogate for the Iron Horses, the engines that powered the transportation of the last century.

The parallel extends into the methods for determining a nation's gross national product: neither transportation of physical products—manufactured or grown—nor electronic transportation of information is directly counted into a nation's total GNP. Yet, in both cases, the convenience, affordability, speed, and accuracy with which the transaction is accomplished directly affects the marketability of products and the viability of the economic system.

If it is difficult to measure the dollar value of communications and computer transfers as part of the GNP, it is even harder to determine what impact international competition and government policy have on that figure. It is generally accepted that foreign competition is a real and growing threat. What is less clear is what effect government has. Its most visible role to date seems to be regulation.

COMMUNICATIONS QUALITY PARTNERSHIP AND VISION NEEDED

What is needed is a more creative relationship between government and these industries. Partnerships between government and the private sector have long been a part of the American tradition. An 1862 Act of Congress led to the transcontinental railroad, generally viewed as having enormous implications for the final borders and economic growth of the United States. The interstate highway program, a partnership between the states and federal government, began in 1956 and greatly facilitated the movement of private goods.

What is sobering in both these examples is the time lag. The congressional railroad legislation came thirty-six years after the first locomotive was licensed; interstate highways came more than fifty years after the first automobile. The first ENIAC computer powered up in 1946. A full scale, visionary government policy would seem to be due.

The new revolution is truly new Society does measure the economic impact of communications on GNP by inference: the amount of money being made by people active in the industry. Where many of the great fortunes accumulated in the early 20th century were based on physical transportation (Vanderbilt in shipping and railroads, Rockefeller in railroads, Ford in automobiles, et cetera) today's, and tomorrow's, fortunes are increasingly based in communications, a communications made possible by computers. When the 1989 *Forbes* magazine list of the richest people in America was published, 70% of the newly added multi-millionaires had made their money in the communications and media field.

The *Forbes* list intermingles the message and the means of transmitting the message. Some of the communications millionaires provide messages; some provide the means to transmit messages; some do both.

Most people accept the coupling of message and medium without difficulty; the startling insight of Marshall McLuhan has become a ho-hum idea. For most of human history such mutual identification would have been unthinkable.

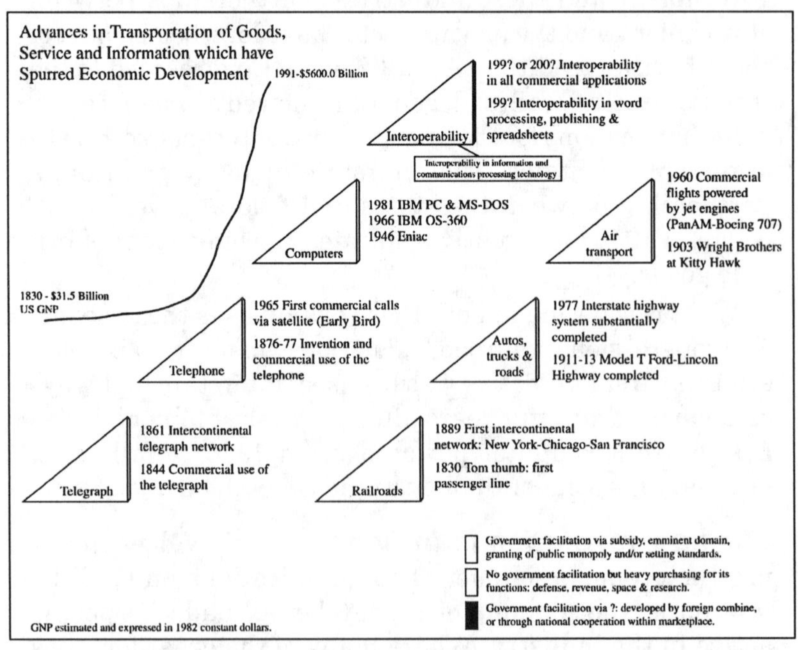

Figure 11.1 Advances in transportation

Prior to the telegraph, the means employed for the transportation of information and the transportation of things were virtually identical. The only difference may have been that a rider carrying a letter could be expected to make better time than a rider carrying a large object. In both cases, however, it was easy to separate the object of the transportation from the means of transport—a distinction that, while clear, still did not always prevent the occasional bearer of bad news from being held responsible for the contents of the message.

In today's world, physical objects can now be carried more quickly than was possible two centuries ago, but the increase in speed is measured in relatively modest multiples. What used to take two weeks to move from point A to point B can, for instance, now be done in two days or, for an extra charge, overnight. In contrast, the increase in speed of moving ideas and information from one location to another has been dazzling; the time is now measured in seconds, or nanoseconds.

TODAY'S EXPECTATIONS

If General Andrew Jackson had had the same access to fax machines that the Chinese students in Beijing had in the spring of 1989, the battle of New Orleans fought after the signing of the treaty which ended the War of 1812 would never have taken place.

In the process of accelerating the speed of the information transfer, the relationship between the means of transportation and the message itself has become more intricate (the computers and procedures driving the transmission of the message have had a growing impact on the accuracy and usefulness of the message itself). This fact alone makes it imperative that both industries focus on quality. While it is not completely true that the medium is the message, it is now more often just to shoot the messenger.

Frustrations There's a built-in frustration for both providers and customers that comes with this new relationship. It was easy to understand how the pony express got a message from one location to another. It probably took little research to find out exactly where the hand-off points were as a single rider switched horses or a message was handed off from one rider to another.

When a message is sent out electronically, who knows how many computers are involved, how many times the message is "handed off" from one computer to another? The average customer neither knows nor cares. How the transmission is accomplished is not a matter of concern; that it is transmitted accurately and quickly, is. When everything is working correctly, the system is invisible.

Risks All of which ups the ante for those responsible for the performance of communications and computer systems. Customers are less able to help identify the source of problems. If a car is hard to stop, the driver can tell the salesman or the service manager to look at the brakes. If a message gets garbled or lost between sender and receiver, the customers external to the system are at a loss. They only know that the system didn't work. And that they suffered as a result.

Expectations... The same principles apply to the communications and computer industries that apply to any human enterprise: the customers' expectations need to be understood and precisely described, and then potential providers need to determine if it is possible to define and fulfill specifications that match those expectations.

If it is possible to exceed stated expectations, then it is necessary, and appropriate, to educate potential customers about the possibilities. No one knew two years ago that they needed to order lunch by fax to save a few minutes, but it is in fact becoming an inalienable right.

Conversely, if it is not currently possible to meet a customer's expectations, it may be necessary to educate the customer to the realities of the situation. A compromise, hopefully temporary, may be in order.

The challenge There are alternative scenarios that are not as optimistic. Customers may be unwilling to wait. Increasingly, they may not have to wait. The international marketplace makes it likely that an alternative to waiting may already be available. And they may find it. Even being the first to develop a product or service is not a guarantee of continued success. Once anybody else, foreign or domestic, can do it cheaper and better, they can move into the market. Once again, quality is a major issue.

Dreams and responsibilities In any case, to assure continuing customer satisfaction in the future it will require a closer relationship with customers that has been the rule in the past. The surveying of customers must probe not only present and past satisfaction, but future expectations and dreams; sometimes unarticulated dreams.

For years, millions of people regretted having to miss a favorite television show because of a scheduling conflict, but there was no loud hue and cry for something like a VCR. Everyone just lived with it. Customers are not in the habit of describing their dreams—or of thinking in terms of their dreams being realized.

There is also increasing awareness that providers have a civic responsibility to insure the safety and the security of the product. Even intangible products must be protected from abuse. Just as liquid-carrying trucks and trains built leak-proof containers so must communications/computer engineers build leak-proof containers for information. Similarly, just as there are procedures and equipment that allow shippers of liquids to safely switch their cargo from trucks to trains to ships, there must be data storage, retrieval, and transfer capabilities and techniques that enable communications/computer operators to accurately move information through the system. If anything, the task is more difficult and more delicate because it is easier to spot corrupted liquid than it is to spot corrupted data.

The more sophisticated customer expectations and industry specifications become, the more every employee of every provider has an all-but-untraceable impact on the final outcome. Employees at every level must be thoroughly trained in the concepts and techniques associated with quality. The question of whether to actually involve everyone in a quality process becomes moot. While there may be some justification for continuing the debate about 100% involvement in manufacturing quality, it must be a virtual given with the computers and communications industries.

A caution and comment With all aspects of quality processes—leadership, participation, measurement, customer satisfaction—extreme care must be exercised to insure that a set of procedures that appears to work at one point in time doesn't become codified as the way to do it. "Management by fact" is stultifying if only yesterday's facts are allowed.

The danger is that as quality professionals become more sure about what constitutes quality in communications and computers they will over-prescribe in such a way as to block any kind of breakthrough or major advance. This is of particular concern in these industries in which the collective intelligence and insight of the people involved often leads to innovation and breakthrough.

One of the uses of a magazine such as this is that it offers a variety of different possibilities for readers to explore possibilities that can break down any barriers self-imposed by an individual's experience to date. Readers can also combine the ideas found here with their own experience, making it possible to transcend what others have done and proved.

The Journal for Quality and Participation, *Jan.–Feb. 1990.*

THE RIGHT QUESTION

"Yes, well, this quality stuff sounds good, but it doesn't really apply to us. We're different." Software units—whether the focus of an organization or a supporting adjunct to the main line of business—are particularly prone to the temptation to avoid quality issues by offering this excuse.

Different in what way? Often the apologist truly feels that his or her organization is "too big" or "too small"—or has a unique, creative operation that would be stifled by an insistence on "doing things right," much less "doing them right the first time."

Robert Townsend (no relation to the coauthor), in his book *Up the Organization* which, in 1970, was ahead of its time, wrote:

> "(Computer technicians) are trying to make it look tough. Not easy. They're building a mystique, a priesthood, their own mumbo-jumbo ritual to keep (their customers) from knowing what they...are doing."

The irony is that software organizations frequently lose the opportunity to utilize their genuine differences. Attempts to ameliorate quality in a programming environment too often most closely resemble the least credible approach, the one devised by old-line manufacturing organizations. An abbreviated history of the quality movement in the United States puts that approach into perspective.

There have been three major types of organized quality efforts in America during the last 40 years. The first was introduced in the years immediately following World War II and was put into place in thousands of companies by top management groups who asked, "Who can we put in charge of this quality thing?"

The result was the creation of the quality control specialists who typically were given some training and a box full of measuring tools, positioned next to the door to the loading dock (or mailroom in the case of paper-and-ideas companies), and told to "stop the bad stuff."

The minute a select few were identified as being responsible for quality, the game was on. The goal of everyone else was to "see if it will get past the quality control guys." If the end of the month approached with shipping quotas unmet, the solution was simple. The quality control guys were sent to another seminar—or, on vacation.

As is obvious by the state of the American economy today, relying solely on such last-chance quality control didn't work in the manufacturing world. It is even less effective—and more costly—in service organizations. As that great American philosopher Lily Tomlin said, "Quality. If it's so good, why do we keep trying to control it?"

The second major attempt at quality in America was typified by quality circles (modeled poorly after Japanese quality control circles). These were formed in response to a new question posed by top management, "Who can we get to be involved in this quality thing?"

And so the call went out for volunteers, from non-management only, of course. Ten percent of the employees raised their hands, desperate to do something to improve the "way we do things around here." Ninety percent, leery of one more program-of-the-month ("Oh, swell, they've been to another seminar!") or productivity push ("Whose job are they after this time?"), chose not to be involved.

Once again, a clear line was drawn between those who had responsibility for quality, for improving things, and those who had none. The fatal, genetic flaws that quality circles carried from birth preordained their quick demise, no matter how spectacular their early gains.

Recently, the top management teams of some organizations, regardless of the size or type of the organization, have begun to ask the correct question: "Who can we afford to exclude from an effort to improve, and to try this quality thing?" The answer is: "Nobody."

This is not to suggest that every person be handed some measuring tools (hardware or software) and be lined up at the door, taking turns catching mistakes.

Beefing up the "correction" effort is not the answer. It is to suggest that every person be enrolled in an effort to prevent errors, inaccuracies, or deviations from occurring. The focus must be on producing what the customer expects and the organi- zation can provide the first time, every time, which is not easy, not quick, and definitely not free. It is a lot of work. So why bother?

Even if meeting customer expectations is low on a software organization's operational priorities ("What do those techno-peasants know about what we do?"), there is one statistic published recently at a National Institute of Standards and Technologies conference that provides ample motivation. To produce one line of finished code in a program in production in America costs $50. To fix one line of code in a program in production in America costs $4,000.

In other words, a programmer who chooses (or is pressured into) "giving it a good shot" with any mistakes to be caught by the quality control guys or fixed "after it goes to Beta Test and we see for sure whether or not it works" is inviting an additional cost of $3,950 for each line of code that does not accurately fulfill a need either defined by the design or expected by the customer. Translating that ratio into time, an hour of time spent up front can prevent the expenditure of two weeks of paid-for effort down the line.

By choosing to mimic the cop-at-the-door/fix-it-when-they-send-it-back approach that has been so costly for America's staggering manufacturing concerns (indeed, it has added to the degree of the stagger), programming units are being wasteful of their owners' money and their people's time. Many companies are willing to accept the hidden cost rather than face the hard-dollar cost of funding a quality process. Quality is an investment. Sometimes it can appear to be a costly investment, but what other investment has a high probability of an 80-to-1 return?

In addition to the dollars invested, quality requires an investment of time. This is often the real stumbling block—especially with top management. Make no mistake, top management will have to devote a great deal of time instituting a quality process. Decisions about how to approach quality require active, informed commitment from the top. A steering committee staffed by corporate decision-makers must define the quality process that is going to work for their company.

A few concepts are fundamental, such as the need to actively enroll every person on the payroll in the effort to improve, the necessity of providing appropriate training, and the correctness and wisdom of saying "thank you" in ways that are heard. Most of all, quality requires a proactive structure that continuously asks, encourages, urges, and allows people to improve whatever it is they do. This structure will need someone whose full-time job is to ensure the process is a success.

A major mistake made by too many organizations that have decided to work toward becoming quality organizations is the hiring of a consulting firm to do it for—not with—them. Consultants must not be brought in as nannies to run the process because top management is "too busy" to be involved in something they want the rest of the employees to believe is very important.

As an example, if Company ABC buys the XYZ Quality System and then hires a phalanx of consultants from XYZ to "install the system," it will always remain the XYZ system, it will never be the ABC system. And the employees will know it from day one. No matter what bells, whistles, or speeches are attached to the XYZ system, employees will always believe that the process is temporary, the result of the boss having had an encounter with a particularly fast-talking salesperson. As a result, the company will never reap benefits anywhere near the potential.

Is there a role for consultants? Yes, consultants can be hired for a defined period to contribute—and leave behind—specific information and/or skills, at a mutually agreeable price. They can be especially helpful during the planning stages for crystallizing issues for the steering committee's consideration. This use of the experience and knowledge of consultants can save the committee hours of research.

The retention by the steering committee of control over the definition of the quality process greatly increases the credibility of the process. In the above example, the fact that the process was defined by a group which included top management of Company ABC tells everyone in ABC two things: (a) top management really is committed to seeing this through and sustaining it (after all, they have already contributed a good share of their most precious assets—time and decision-making authority), and (b) it is a process that will truly fit ABC.

The steering committee defines everything from inaugural ceremonies to precise mechanics that will actually enroll all employees from CEO to newest hire, fromtraining to the program for recognition, gratitude, and celebration. And it designates a full-time director of the process (with an appropriate staff) to assure its sustainability.

This approach, the third type of quality process described earlier, is particularly appropriate for software operations. Since it relies on pushing authority down to the level of responsibility, it focuses on preventing problems rather than foisting responsibility to "stop the bad stuff" onto a small group of beleaguered quality control specialists.

In the near future, software is scheduled to be the target of a concerted competitive effort from Japan. The software organizations that begin soon to solve quality issues effectively and produce programs that live up to expectations for quality will succeed. The others will fail. It is as simple as that. Alibis won't serve; nor will half-hearted efforts.

It is imperative that American software firms not go down the same paths defined and traveled by their manufacturing brethren over the last few decades. Disaster lies down those paths. Alternatives exist—alternatives which thrive on the uniqueness of each organization.

Quality Data Processing, *Jan. 1989*.

TOTAL SERVICE QUALITY

In an interview in October 1988, Kent Sterrett of Florida Power and Light labelled service quality in the United States as "primitive" when compared to techniques employed to assure quality in manufacturing. The remark suggests an interesting perspective, not entirely in keeping with the experience of many American consumers.

While it is true that formal efforts to improve quality of service are often tenuous, the chances in America of getting a good hamburger, politely served in a clean environment remain higher than the odds of getting an acceptable American-made car.

Certain facts in the "American history of quality" are, however, undebatable:

- The concern for quality first appeared in the manufacturing segment of the economy.
- Statistical tools for quality control were developed in a manufacturing environment.
- The service segment is now realizing that quality is a major concern for them as well.

Does manufacturing offer a model for service quality? The answer is yes and no. Quality efforts in a manufacturing environment revolve around measurement. The logic of the situation seems to be this: in manufacturing, the most expensive ingredients are machines. Through quality assurance it is possible to measure the performance of the machines and to work at making the correct mechanical adjustments to minimize the impact of the inherent randomness. A well-cared-for machine has predictable behavior and a predictable life span. Using that predictability, it is possible to maximize productivity and minimize cost.

Of course, this overstates the case, but it does indicate the main difficulty in applying manufacturing techniques to a service environment. In a service organization, the most expensive ingredients are people. The range of predictability widens considerably. Machines can't quit and go to school; machines can't be hung-over; machines don't have personality clashes. People can and do.

NEW ORIENTATION NEEDED

From the outset, quality in the service world requires a different orientation than quality in manufacturing. Service must give priority to leadership and human resource utilization. It is an orientation likely to make many statistically-oriented quality professionals uneasy, based as it is on something difficult to control and measure: attitude.

Technology and measurement are necessary in the service sector, but they will not have significant impact on service quality without the underpinnings of employee cooperation and understanding from everybody.

Any successful effort to improve quality, whether in manufacturing or service, will have to address *people* issues at some stage. It is possible in the manufacturing world, however, to conduct a highly successful quality process actually involving only a tiny portion of the people on the payroll. The effort can also be fundamentally top-down. Top management need not share authority, but just name a few action teams to study detailed problems for final top management review. After a plan is developed, an extensive information program lets people know why some changes are being made and, hopefully, encourages cooperation with anyone identifying themselves as a member of an action team. Short of sabotage, the rest of the employees can pursue business as usual.

Service quality is bottom-up That won't work in service. Everyone must assume responsibility for their own actions, and everyone must cooperate. Because the percentage of employees who deal with "real world customers" is considerably higher in service than in the average manufacturing operation, one surly clerk on the telephone can undo a management master plan of heroic proportions. Quality is fundamentally bottom-up.

Service quality is not only harder to establish, it is far more difficult to sustain. Management is no longer good enough; leadership is necessary. It will not, for instance, be possible to micro-manage changes or improvements. If everyone becomes involved, it will be impossible to filter every decision through one central point for approval prior to implementation. Everyone will have to share authority as well as responsibility. Management at every level must trust those who work with and for them—and must return the authority they have slowly usurped over the years from their subordinates.

Leadership, authority, cooperation, responsibility — all of these are difficult to produce and difficult to assess. The payoff for those organizations that make the effort is that well-led employees in a service organization, with a quality focus, can make sweeping changes in what they do, and in how they do it—far more quickly than in technology-bound manufacturing.

TEAMS AND SUPPORT SYSTEMS

The question is how to create a corporate culture that institutionalizes these attitudes. What is needed is a proactive structure that positively encourages change. Quality teams composed of every employee from the CEO to the newest hire (working on problems defined by the team itself and authorized to make changes in their area of responsibility) have worked in a number of companies.

These teams must be supported by a network of communications, training and recognition. Having correct information and knowing how to act on it are essential. Too often, information is furnished without the latter—"quality by exhortation" programs fall into this category. Training in the concepts of quality, in helping unit leaders run participative meetings, and in problem solving and analysis enable team members to act. (When in doubt follow Tom Peters' rule of thumb: if things are going well, double your training budget; if they aren't going well, quadruple it.) "Thank you" must be said, and it must be said in a variety of ways to insure that every deserving person *hears* it.

Measurement support No quality process is complete—whether service or manufacturing—without measurement. The cautionary note for service is to not let the measurements drive the effort, but to use the measurements as tools. Appropriate measures must be taken both as a source of ideas and as a yardstick of progress. Don't chart or graph things simply to please top management or the quality experts. Sophisticated measures are not always the answer. It is useless to measure the number of smiles per hour per flight attendant. Attitude goes beyond mechanics. It is helpful, however, to know who receives praise and who is the object of complaints from passengers.

A service organization that wants to earn and keep a reputation as a quality organization—that wants to increase its profit margin while pleasing both its employees and its clients—should start by evaluating manufacturing's experience with a critical eye. The fact that the manufacturing world has spent more time and money on the topic of quality than has the service world does not mean that it has a corner on the expertise... or even that manufacturing understands the problems facing service companies. Measurements will most likely not be as neat—or as easy to take—in service as in manufacturing, but manufacturing has developed tools which can be adapted to the service environment. Value analysis is such a tool and deserves thorough consideration.

COMMON SENSE AND THE FUTURE

Common sense should play a large part in deciding how to address quality issues in the service sector. Service quality processes should be built from the ground up to fit individual organizations. The attempt to graft a manufacturing–based, one-size-fits-all, process onto a service organization is a waste of time and money.

The stakes are high. American manufacturing during the last two decades has seen its share of the economy slip to approximately 20 percent, despite legions of quality professionals who have tried to stem losses from foreign competition and changing market conditions. Between 1970 and 1985, America experienced 3,727 plant closures and the loss of approximately 22 million manufacturing jobs.

Apologist academicians who speak of a maturing national economy as the reason for the diminishing of manufacturing's contribution to the total are not living in the real world. If that judgment were true, the United States would not already be running an international deficit on the service side of the ledger, too. The deficit is there, and it continues to grow.

If service does not right itself and, in the process, buy time and provide leadership and examples for manufacturing, America is likely to give up a large degree of control of its economic destiny. By pioneering new approaches in quality, however, service can succeed where manufacturing failed as America begins the 21st century.

Journal for Quality and Participation, *Volume 12.1.*

TOTAL QUALITY LEADERSHIP OR PARTIAL QUALITY MANAGEMENT?

Think of the adjectives normally used by Marines to describe leadership: bold, daring, inspirational, innovative, effective. Then think of the adjectives normally used by Marines to describe management: stuffy, slow, multi-layered, bureaucratic, over-controlling. Now consider the adjectives that more accurately describe the Total Quality Leadership (TQL) initiatives currently being imposed on various Marine commands. Is it any wonder that Marines all over the world are digging in their heels and muttering, "Not on my watch"?

The concept of leadership ranks only behind patriotism and honor in a Marine's hierarchy of values. As a result, Marines instinctively rebel against anything that assumes the name without openly (and passionately) playing the game. Embracing the characteristics taught in boot camp or The Basic School as inherent to leadership would solve part of the problem, but even that may not be enough to win support. Older Marines remember previous large-scale attempts to transfer civilian management practices to the military. Younger Marines have heard embellished stories about the failures that followed.

This first occurred in 1961 when Robert S. McNamara became President Kennedy's Secretary of Defense. He came to his cabinet position from the Ford Motor Company, where he was a top executive, and he brought with him Ford's management practices. Degrees in management suddenly had value in the military. Economics of scale brought similar uniform items to the various services. Full participation in society brought in tens of thousands of new recruits specializing in economics and quantitative management. Thanks in no small part to these newly learned management skills, the Vietnam War was fought every bit as wisely and successfully as American automobile manufacturers fought subsequent battles against Asian competition.

In the 1970s, it was HumRel, or human relations. The idea itself was sound, but poorly presented and explained. As a result, it was viewed with suspicion throughout the Corps. The objection: How could a civilian world that obviously knew so little about race relations teach the Marine Corps anything? This position was arguably true. Citing the demonstrations—violent and non-violent—of those years, it was easy for Marines to conclude that the Corps was already way ahead. "We may not be perfect," said the Marines, "But we're better than the civilians are."

Throughout the 1980s, however, the civilian world was quietly undergoing a revolution in the name of quality, and TQL was the logical extension. Unfortunately, TQL received a questioning, although understandable, reception in many commands. But TQL is different. It is not a clump of management practices to be slavishly imitated. It is not HumRel revisited—at least, it doesn't have to be. Quality and the Marine Corps are a natural pairing. The Marine Corps is the perfect place to initiate a quality process because Marines have always prided themselves on their ability to do more with less, and efficiency combined with effectiveness is one of the aims of a quality process. In addition, Marines everywhere think of themselves as leaders and a quality process can flourish only where there are leaders.

Still, Marines have hesitated to get behind TQL, and they have been right. Both the vocabulary and the methodology currently in use lead them to believe that something alien is being superimposed upon them. TQL borrows heavily from the manufacturing industry, and the Corps more closely resembles the service sector, which isn't easily measured in terms of output or production efficiency. It further differs from manufacturing quality in that it is more complex. Civilians have only recently begun to wrestle with this reality, and the Marine Corps has a golden opportunity to make a significant contribution here, particularly if it can avoid the pitfalls that plague so many civilian quality processes. Unfortunately, to date, it doesn't seem to have done so.

Take the word Total, for example, as in "Total" Quality Management. Too many leading civilian gurus (and apparently the Federal Quality Institute and the Navy and Marine Corps TQL gurus) use Total Quality to describe the efforts of a few teams—project action teams or quality management boards—made up of relatively senior people working on problems either chosen by the senior people in the organization or, at the least, approved by them. When the resolution of these problems, to one degree or another, impacts everybody in the organization, the results are deemed Total.

Wrong. Quality has two opportunities for improvement: top down and bottom up. It is not enough to make decisions that impact everyone. To rightfully use the word Total, a quality process also has to offer every person in the organization an opportunity to impact decisions. Only an approach that combines sweeping changes made by the few with a structure that facilitates incremental changes in every job at every level covers all the bases.

The litmus test of a quality process and its claim to be total is this: Whose behavior is being changed as a result of the decisions that are being made? If decision-makers are always changing someone else's behavior but never their own, the claim that Total Quality improvements are underway is nonsense. Quite simply, the phrase Total Quality cannot be used with any integrity unless every single person in the organization is literally, formally, and continuously enrolled in the effort to improve.

Is this possible? Yes. A growing number of civilian organizations are doing it. One, the Paul Revere Insurance Group, has a successful process in place. It was defined over 8 years ago with military leadership principles very much a part of the discussions and definitions.

When Paul Revere applied for the Malcolm Baldrige United States National Quality Award in 1988, the cover letter that went with the application ended with the statement that, "You can walk into the Paul Revere building, walk up to any employee, ask him or her what Quality Team they are on and what they personally are doing about quality this week and get answers to both questions. And that is revolutionary." The company was one of two finalists in the service category that first year.

An organization can only support total involvement if it has a clear understanding of the word Quality, a definition that crosses all functional and structural barriers. During the last decade, older definitions from manufacturing that seemed to ignore the customer have been replaced with newer customer-focused ones. Quality is no longer simply conformance to the requirements of management or fitness for use, instead it has become synonymous with the idea that the vendor should delight the customer.

While an important change, all these definitions ignore the tension between the legitimate constraints of an organization and a customer's wish list. That tension exists and always will. Paul Revere developed a definition of quality that acknowledges this tension by introducing three separate elements: quality in fact, quality in perception, and customer satisfaction. The result is more that a slogan, it is a definition that is imminently practical for day-to-day use.

Quality in fact is reached, or achieved, when someone does exactly what they intend to do. That is, they meet their specifications. They do their job exactly as they understand it—using the resources at hand, they conform to the requirements of management. While it is necessary, it is not enough—what one Marine is doing (in good faith) may not be what the next Marine wants, needs, or expects.

Quality in perception can be correctly claimed when someone else believes that the service or product being offered is going to do or to be what they want. Purchases and plans are based on quality in perception; choices are made based on what the recipient expects to receive. But what if the recipient is subsequently disappointed? Mutual trust and future effectiveness suffer.

Only when quality in fact matches quality in perception is the customer satisfied. While both quality in fact and quality in perception are necessary for success, it is possible to work on one or the other independently.

One more word needs to be defined to make the vocabulary usable, that is the customer, who is anyone to whom a person, unit, or organization provides a service, product, or information. Notice that this definition does not equate a customer with a consumer. And everyone has customers, even if no money is being exchanged. The Marine Corps has chosen the term "end user" instead of customer. Either term works as long as the above definition is adopted.

Stripping off all this verbiage, a quality process is a continuous, proactive effort to determine a customer's expectations, compare them to one's specifications, and see if there is a match. If the match is not exact, something needs to be done. The first choice involves changing the specifications to match expectations, e.g., "Oh, that's what you want! I can do that." The other choice involves lessening the customer's expectations through education, e.g., "I can't possibly have it all here by 1400 Thursday, but I can have two-thirds of it here by 1100 on Thursday and the rest by 1630—would that work?" No mind reading or second guessing the customer is allowed.

Once specifications match expectations, however, the task becomes straightforward: perform. Be aware that one of the things that make service quality more difficult than manufacturing quality is the rapidity with which expectations change. There must be a frequent effort to verify expectations because they are going to change. Count on it.

Total Quality requires leadership. It is impossible to manage anyone into quality. Quality is not something you do to someone or for someone (hence the limitations of the top-down approach), it is something you do with someone. In order to establish a quality process in an organization, the beginning point is trust. The highest ranking members of the organization must believe that everyone in the organization shares a common goal, and that they are capable of achieving it.

Leaders have no trouble with this starting point. Managers, however, get nervous. What's the difference between the two? A manager cares that a job gets done. So does a leader, except he also cares about the people who do the job. In other words, a manager is a leader who hasn't finished evolving yet.

All that a quality process does is give leaders a structure that institutionalizes their gains. No longer will every new commander have to start all over again; the improvements in the organization will not leave with the departing leader. Additionally, there is no need to start a quality process from scratch either. Don't reinvent the wheel. Civilian experience is not entirely without merit. Borrow liberally from it, just remember who and what you are.

What follows is a discussion of seven principles that can serve as the underpinning of a quality process. If they look an awful lot like small unit leadership, they are. Quality and small unit leadership have far more similarities than differences. Taken together, these principles form the foundation upon which a TQL revolution can transform your organization.

- The first principle involves *top management commitment,* meaning that all senior officers and staff noncommissioned officers in a given unit are committed to the implementation of the quality process. This calls for more than lip service, it calls for action.

This commitment to quality will have to be demonstrated repeatedly. One of the best and most convincing ways to do this is by actually changing and improving what leaders themselves do. In other words, instead of the senior members of the command making changes in what subordinates do, they must make changes in what they themselves do; they must first improve their own contribution to the unit. This activity must not be kept secret, even though it does carry with it the admission that senior Marines are imperfect, a public admission of something that most junior Marines are all too readily aware of.

Besides being active and obvious, the commitment of the leaders of a unit that is introducing and/or practicing a quality process must be informed on the topic. While they should certainly read Dr. W. Edwards Deming's *Out of the Crisis*, and absolutely must understand the underlying message of that book, they should not stop there. Books such as Tom Peters *In Search of Excellence* and *Thriving on Chaos* and Masaaki Imai's *Kaizen* are invaluable additions to a professional library. Of course, *Commit to Quality* and *Quality in Action* (which draws specifically from several military examples) by the authors of this article are also highly recommended.

Another excellent—and necessary—way to demonstrate commitment to the implementation and evolution of a continuous improvement process is by creating the broadest possible structure for improvement. It all depends on the first question asked by senior officials of an organization. If a leader responsible for TQL implementation asks, "Who can we make responsible for this quality thing?" it is destined to fall far short of potential. If, however, a leader asks, "Who can we afford to leave out?" the obvious answer of "Nobody" will set the unit well on the way to Total Quality improvements.

Involving everybody inevitably leads to the idea of empowerment. This is a term that has gotten a lot of attention over the last 3 years, with several books devoted solely to it. Marines will recognize it as something that was the cornerstone of their early leadership courses: authority equal to responsibility. But it is not some fuzzy, do-your-own-thing, rag-tag system of operating. Empowerment means that if someone is going to get blamed for something (i.e., they are responsible for it), then they should be given the authority to change it. Consequently, teams of Marines who have ideas on how to improve procedures that they are responsible for should not have to play "mother may I" games. It should be understood that the concept of authority equal to responsibility is alive and well within the Marine Corps.

- *Leadership* is the second principle necessary for a successful quality process. Basic leadership skills already in place in the Corps make a quality process relatively easy to understand and assimilate. And the introduction of a true quality process will enhance the practice of leadership by bringing new emphasis to the theories and practice of leadership. If properly nurtured, a quality process will reopen long-forgotten, highly valuable discussions, which should remind senior Marines that the Corps is composed of adults who share a common dream and can be trusted. Juniors will be reminded that the freedom to act responsibly—to grab the initiative and practice appropriate authority—is more than a vague theory, it is expected and necessary. Initiative at every level, to make what is within reach as good as it can be as quickly as possible, is as natural a feature of a quality process as it is of effective leadership.
- The third of the seven principles involves *100 percent involvement—with a structure.* It is vital that every person be actively and continuously involved in TQL. It is equally important that this involves establishing a framework that makes it not only possible, but easy, to contribute. To simply announce, "All you people should work smarter, not harder" or "We're going to do this quality thing, like it or not," won't suffice. There must be a recognizable, understandable system that is bolstered and given life by a passion to improve. There are a number of ways to do this but any structure must deal with two questions: "Are we doing things right," and "Are we doing the right things?"

One of the ways to address the first question is to form

groupings, i.e., quality teams that meet on a regular basis to decide how to perform their jobs more efficiently. If this path is chosen, every single person should be on one of these teams. The teams will set their own agenda, make changes as needed, and report their progress to the entire unit so it can benefit from them.

A network of teams that embraces an entire command might seem like an adminstrative nightmare. Not necessarily. Since the teams automatically have authority equal to their responsibility, there should be little need for any paperwork. In fact, compared with the currently proposed system of quality management boards and project action teams, it is an all-inclusive, virtually self-managing system.

The other question that needs to be addressed. "Are we doing the right things?" can be answered through a variety of methods. Techniques such as value analysis, process analysis, and blueprinting can all help, but they must be made by those Marines responsible for implementing TQL initiatives.

- Principle number four involves communications. Again, the Marine Corps has long acknowledged that information must be pushed throughout its organization. Hoarding information—in the belief that knowledge is power and exclusive knowledge is extraordinary power—is contrary to the basic notions of both leadership and quality.

Actively seeking information and advice from junior officers gets them involved in the decision-making process. Delegating some decisions as appropriate should also occur more often. Remember that there are three styles of leadership: authoritarian, participative, and delegative. Authoritarian leadership is still a valid option under certain circumstances. The ability to employ authoritarian leadership and get an absolutely instant response, for instance, is vital in combat. All Marines know that. They also know when they are not at war. If everything is a crisis, nothing is a crisis.

Seniors must routinely, consciously, listen down so that they can benefit from the lessons learned by their juniors. But listening down should not be demeaning to a boss—it's an important facet of leadership. And any boss who has ever said, "I know the jobs of all my people," is lying both to the listener and to himself—unless, of course, he or she has wasted a truly incredible amount of time.

- *Training* is the fifth principle. Quality-specific training is absolutely essential—but not excessive training. Too much TQL—however valuable—is almost as bad as no TQL. All that is truly needed is a short, several hour-long orientation course for everyone in the organization and some prior, detailed training for the key people who will be running the program. Look at it like this. If 10 people (the normal size of a quality team) are gathered in a room, in good faith, with the intent to try to figure out some ways to make things better, how many need to know how to run a meeting in a participative, efficient manner? One. And, how many need to have a good handle on the various problem-solving tools and other mechanics for guiding a discussion towards a solution? Again, one—the same one. With the passage of time, the other nine will soon pick up a good share of information.
- The sixth principle involves *measurement*. Everyone has been burned too often by some measure or another to feel completely comfortable with the idea of measurement. In addition,it is all-too-easy to declare that, "You can't really measure what we're doing." Nonsense. Everything can be measured. The key to making successful use of measurement is to be sure that the data generated and gathered does not become either a religion or a weapon. Measurements should be used as a source of ideas and as a way to track progress. One example of the latter would be a Quality Idea Track-

ing Program that could track the progress of an organi- zation's quality teams, either in terms of raw numbers of ideas or of dollars saved by those ideas.

One common form of measurement is the use of surveys. Surveys are generally easy to create and provide immediate feedback. One survey that has proved successful in the past involves giving each member of an organization a sheet of paper with this statement printed across the top: "The main thing that prevents me from doing my job right the first time is..." The recipients of the survey should then be given a couple of days to fill it out. Warning: this should not be tried unless the leaders of the organization are prepared to read the answers and begin changing.

- The seventh and last principle upon which to build a firm TQL base involves *recognition, gratitude, and celebration.* Essentially, what this principle boils down to is thanking those underneath you. Why should one say thank you? Isn't that their job? This is not true even in an objective sense. "Their job" was defined by the last paycheck they received. If there were no sanctions or warnings, "their job" is whatever they did—or got by with—during their last pay period. It follows that if they made an improvement on that performance, they deserved to be thanked. The good news is that by thanking them, "their job" is redefined at this new, higher level.

In building a program of recognition, gratitude, and celebration, it must be remembered that different people hear thank you in different ways. The reward that gets one person excited leaves the next person cold, which partially explains why some Marines always wear all of their ribbons and others rarely do.

For some people, a material thank you conveys the message. For others, something symbolic connotes thank you. For still others, public recognition; and for still others, a personal thank you by the senior person. Making all of this trickier is the fact that people themselves often cannot tell in advance which thank you is going to work for them. The only solution: say thank you in several ways to each deserving person and let him or her hear the one they want.

Celebration is also a natural part of a quality process, just as celebration has always been an integral part of any successful unit's culture and habits. Stop and party. Remind each other how good you are; enjoy each other.

None of this is nuclear physics, and all of it is well within the reach of any Marine Crops unit—and it shouldn't take forever to get it going. Too many are still hiding behind the statement that "this quality thing will take 8 to 10 years." It does take several years to get a fully mature quality process—but that is several years of continuous, active, evolutionary effort, not several years of pushing things off to one's successor.

For any unit of less than 2,500 people, there is no defendable reason to not have a well-defined, all-inclusive, quality process up and operating within 6 months of the decision to get serious. And while it may take several years to achieve complete success, a unit should begin seeing some improvements right away.

If Marine units are not willing to make that commitment and buy into the TQL concept of having everyone involved in an organization contribute to the continuous improvement of all facets of the organization, the word Partial should replace Total in Marines' lexicon, for only a few Marines will be participating in this revolution, and it will fail, at least with regard to the Corps.

Marine Corps Gazette, *March 1993.*

5

FINAL THOUGHTS

Ideas for Consideration
Keeping Quality Alive
Putting the Focus in the Right Places

Quality has proven its worth. Customers insist on it and fortunes have been made by those who have listened to their customers. Yet, it must be admitted that the reputation of "quality" has taken some beatings—at least in America and Europe—in recent years. This is due in large part to the large bills-for-services-performed presented by "Quality Consultants" for programs that were shortsighted at best. But the fault has not been with the concept of quality, the fault has been with misguided practitioners and their uninformed (or, at least, under-informed) tutors.

The definition of a quality process takes time, mostly the time of very senior executives in the organization. The initiation and maintenance of a quality process takes even more time and more resources. Is it really worth it? It's a valid question and one that Motorola asked itself a few years after it won the Baldrige in 1988. They were, at that point, spending several million dollars a year on quality-specific training alone.

To answer the question, Motorola hired an outside firm to audit its own quality-specific training and the measurable results of that training. After several months, the consultant firm provided Motorola with proof that the return on investment that it was getting from its quality-specific training was 30-to-1. That is, for every dollar they invested, they received thirty dollars in quantifiable benefit. Quality makes money, when *done* correctly.

In 1994, a national study determining a Customer Satisfaction Index (CSI) for approximately 200 companies in the United States was launched, used extensive one-on-one interviewing as the data-gathering device. It has been shown that for the companies that show a positive change in their CSI between two successive years, there is an average increase of profits (measured as return on assets) of 10 percent. For those companies who evidenced a decrease in their CSI, profits fell an average of 14 percent. Not only does quality make money, those companies that go backward lose money. No one gets to stand still; you either win or you lose.

This final section presents a potpourri of ideas on enjoying work, transformation, followership (yes, you read that right), and creativity, followed by a number of ways to check your progress and keep things on track.

12

IDEAS FOR CONSIDERATION

A better understanding of the dynamics of a workplace leads inevitably to a better understanding of how to maintain a vital quality process. For one thing, involvement in a quality process can be a source of joy for employees. They will, after all, have more control over the decisions that determine what they do, hour by hour. They will be able to take part in insuring that they can be proud of what they do. People will be saying thank you to them. In short, the pairing of "Work and Enjoyment" is a natural one.

Introducing a quality process successfully will change any organization forever. Once the employees are used to being treated as full partners in the evolution of the company, they will not willingly give up this role. Once customers are accustomed to receiving quality goods and/or services, they will turn to another provider rather than accept less. What happens to the organization during the implementation of a quality process is correctly called a transformation. "The Role of Leadership: Examples from the Insurance Sector and US Marine Corps" discusses this aspect of a quality evolution, drawing from two sources already seen in this book: the Paul Revere Insurance Group and the American military.

There is no question that leadership is important and hundreds of volumes have been written on leadership in virtually every language heard on the planet. But what about "followership"? Although not found in most dictionaries, as described in "Followership: An Essential Element of Leadership," the concept is every bit as valid. Every person in an organization spends some part of his or her day functioning as a follower. Here too, the military is a primary source of benchmark information for this very important and honorable role.

"Creating More Creativity" explores how to encourage creativity. Whether leaders or followers, creativity is an unavoidable by-product of a solid quality process that reaches into every corner of an organization. It may be hard on the nerves, but it's good for the bottom line.

WORK AND ENJOYMENT

Removing barriers to pride, not just for the hourly worker but for employees at all levels, will open the doors to another feeling as well: joy.

People who have cause to take pride in their work, who are able to express and demonstrate that pride, will have reason to look forward to coming to work and to take enjoyment from what they accomplish.

Be happy at your work. —Mao Tse Tung.

Remove barriers that stand between the hourly worker and his right to pride of workmanship.—Dr. W. Edwards Deming (Point # 11)

He can't be working very hard... he seems too happy.—An anonymous capitalist executive.

When put as directly as in the third quote, the idea as stated seems foolish, yet employees are often counseled or criticized—or, at least, looked at strangely—if they are too blatantly happy during work hours. American businesses have, unfortunately, peopled their senior ranks with too many men and women who believe that having a good time and doing a good job are incompatible achievements. Laughter has, as a result, become a rare thing in too many offices and organizations. In insuring that everyone takes things as glumly as they do, joyless executives have constructed tremendous barriers.

Perhaps the collective hesitancy of so many Americans to allow themselves to be obviously pleased to be doing something they value, and which is a source of pride and satisfaction for them, is a hangover from the traditional American Puritan work ethic. If so, it is one tradition that the country can afford to drop.

It is granted that some employees—at all levels—don't take their responsibilities seriously enough at all times and should be called to task for their errant attitudes.

Care must be taken, however, to not confuse giddiness with a sense of humor or *not being serious enough* with a sense of proportion.

The value of humor and a sense of proportion These two qualities—a sense of humor and a sense of proportion—are so precious that they should be nurtured carefully when discovered in a subordinate and appreciated deeply when evident in a senior.

Seriousness and humor are not, after all, mutually exclusive ideas. In fact, they are complementary—witness the fame of the cartoon strip Doonesbury or the writing and commentary of Mark Russell. Without a sense of humor and a sense of proportion, there are no gradations of seriousness. Everything becomes an equally tense-jawed crisis. And, when a *real* emergency develops, the story of the boy crying "Wolf!" is replayed.

Some executives will say that their personal reward comes in a quiet feeling of satisfaction and that the same should be true for their employees. That's nice—but not everybody responds that way. Some people need access to an outward, visible expression of their satisfaction. Why shouldn't meeting or exceeding goals make a person happy? And why shouldn't happiness with its accompanying features of smiling and laughing be part of the daily picture?

On some days, funny just happens... The inescapable fact is that people at all levels of management and non-management do genuinely funny things from time to time—not intentionally funny, perhaps, but very funny nonetheless. Not to acknowledge the humor in an obviously humor-filled situation is to signal to those around you, junior and senior, that all business is to be considered grim.

This is not to say that everyone should always have a song in their hearts, a smile on their lips, and a joke on the tips of their tongues. But people should enjoy what they do—work should be a source of pride, pleasure and, when appropriate, humor.

On other days, boring or serious dominates... Admittedly, some days are irretrievably boring. And on some days, one person's individual impact on another—or an organization's collective impact—is deadly serious. Even if the impact is limited to influencing someone else's mood for the rest of the day, that impact is quite real and, to the affected person, very important. That interdependence injects enough tension into everything that people do without management having to add other peripheral, and frequently artificial, pressures.

There is a practical side to all this Someone who enjoys his or her job, for whom going to work is not a spirit-breaking experience, is going to do a better job. If they are allowed, encouraged and/or otherwise feel comfortable enough to openly take pride from their accomplishments, to appreciate the inevitable humor in their surroundings, and to have an honest appreciation for the relative importance of their various tasks, they will work more efficiently and more effectively.

Leaders and managers The name of the person who coined the timeworn phrase "*A subordinate isn't happy unless he or she is complaining,*" has been lost. This is a kindness, a generous decision on the part of Clio, the goddess of history, since the speaker was obviously a poor leader awash in self-justification. But then, he or she may simply have been a *manager* who had forgotten or never learned how to *lead.* He or she cared only that the job got done. A leader cares both that the job gets done and that the people doing the job are properly provided for and thus have little reason to complain.

For those unused to enjoying themselves at work...

If the manager in question is out of the habit of enjoying himself or herself at work but wants to change, memorizing a joke or two may well be a place to start. The jokes don't have to be *world-beaters*; the fact that he or she works a joke into the conversation (or simply announces, "I have a joke to tell") will be a signal to all that humor is not anathema. An even more basic beginning might be to place–in an obvious spot–a sign saying something like, "To have a good time at your job, no matter what it is, is a sign of intelligence."

What can I do? How can a manager bring some humor and enjoyment to his or her organization—or his or her piece of the organization?

Step one... Start by being visible, both during business hours (Tom Peters' timeless advice to "manage by wandering around" applies) and after hours (for example, are there any intra– or inter–mural sports events involving the manager's subordinates?).

Step two... The next step is to simply be available—and ready to enjoy the moment.

A manager who makes the effort to connect with his or her subordinates on both an intellectual and emotional level will grow in both stature and perceived approachability. What he or she learns about those subordinates will by itself be worth the effort. In addition, what the manager learns about how work is actually done on a day-to-day basis will help him or her follow up on other of Dr. Deming's *Fourteen Points*—such as "Find problems," "Drive out fear," and "Create constancy of purpose." But, most of all, the manager and the folks who work for and with him or her will have fun. The manager's sense of humor and proportion will be evident to all. Hierarchial position is no bar to enjoying each other's company and any group of people who know and enjoy each other's company is going to be a more productive workforce.

Your bonus... A bonus benefit to all this is that both the manager and those who work with and for him or her will go home in better shape to face their families. It's a delight-filled cycle.

Being a member of a thriving company should be a source of great pride and happiness Many companies have come through rough times, when the beat of a very serious drummer set the tone for the organization. A more joyous beat is needed as the 21st century looms ever larger: something from John Phillip Sousa, perhaps?

Final Thought

Any organization that persists in the practice of subtly or openly punishing its employees for taking pride in their work and enjoying themselves in the process is cutting itself off from its most valuable asset—the enthusiastic cooperation of its employees in the continual improvement of the organization. Soon the firm won't have to worry about the 21st century.

A word on Mao's quote It should, of course, be admitted that the quote from Mao Tse Tung at the beginning of this article is taken out of historical context. Mao's idea was certainly commendable but his execution (to include the execution of countless workers) left a great deal to be desired. All too often throughout history, a dictator's hips and lips have been seen to go in different directions.

Journal for Quality and Participation, *Dec. 1995.*

THE ROLE OF LEADERSHIP: EXAMPLES FROM THE INSURANCE SECTOR AND THE U.S. MARINE CORPS

It sounds so intimidating, so final, to use the word *transformation* rather than simply *change*: rather like the difference between a compound and a mixture that was memorized in chemistry class decades ago. Anything that sounds so permanent promises often to be quite dramatic when it actually happens, much like the minor explosions that were the result in the chemistry lab when the formation of certain compounds was attempted.

While it is possible to have dramatic (and usually tragic at the same time) instances of organizational transformation, the ideal transformation for an organization is a far less stressful exercise and it should take place on a continual basis rather than being keyed to a single, searing event or to a series of crises. Such a transformation will require leaders at every level, but particularly at the upper echelons of the organization, and it will require an understanding of the concept of intentional transformation through the acts of an intellectual embracing of an idea, joined with the emotional acceptance of a goal and the means of achieving it.

Consider the case of the Paul Revere Insurance Group in Worcester, Massachusetts in the mid- to late-1980s. The senior management of that organization made the conscious decision to pursue the goal of becoming a quality organization. The basic motivation was simple. It had become evident to them that an organization that was perceived to offer quality products and services, and that could be seen to truly meet those expectations, stood to make a great deal of money.

To their lasting credit, the senior managers also recognized that "quality by proclamation" would not work. That is, it wouldn't—it couldn't—simply be a matter of declaring the company to be a topnotch organization and issuing a new set of standards for employees to meet, and then waiting for the money to roll in. Specific steps would need to be defined and, in order for these steps to be followed, the employees at all levels would have to be convinced that the senior managers' hips and lips were moving in the same direction.

It is important to note Paul Revere was, even at that time, a successful company by any standards, and a leader in its chosen field of individual disability income insurance. However, despite a track record of innovative products and of service to its community, it was a tradition-bound company in which the relationship between layers of the organization was clearly defined: communications flowed down, not up; each portion of the company did its defined job pretty much in isolation. If the aim of freeing up the communications was to succeed—and thus allow ideas from all levels to become part of the organization's effort to improve—there would need to be some fundamental changes in the way people interacted across bureaucratic lines, both horizontal and vertical.

To address this, the director of the company's Quality Has Value (QHV) process—and co-author of this article—devised a program titled 'Program for Ensuring that Everybody's Thanked' or PEET. It was an attempt to institutionalize an idea that was getting a good deal of publicity at the time, MBWA (Management By Wandering Around). The phrases and the practice had been coined at Hewlett-Packard but had then been broadcast to the business world by Tom Peters in his book *In Search of Excellence* and in his subsequent speeches.

The idea was simple enough: managers should spend part of each day away from their desks, talking with "real people," the folks who worked for them and who did what Peters referred to as 'honest work'. It's an easy idea to nod one's head to in agreement, but very tough to actually make part of one's everyday habits. There are, after all, things awaiting attention in the 'in tray'; those "honest workers" probably don't want to be disturbed; it's tough to think of something to say to a stranger—the excuses are many.

To make it work, to begin the transformation, required some specific procedures. The mechanics of the PEET program began with the delivery of a PEET sheet to the desks of each of the top 24 managers in the company on the first of each month. Each PEET sheet contained the names of two Quality Team Leaders (at Paul Revere, every one of the 1,250 employees was on a Quality Team, so there were approximately 125 Quality Team Leaders, with some turnover virtually every month). The individuals on a particular senior manager's PEET sheet may or may not have worked in his or her chain of command and could be at any level of the organization. The senior manager's task? To, at some time during the month, seek out the two people on his or her PEET sheet (one at a time, unless the two were co-leaders of a single Quality Team) and talk to them, on their turf. The topic of the conversation was left unspecified but, if all else failed, they were encouraged to talk about the quality process.

Intellectually, it was an easy program to accept and agree to. The acceptance of the President of the company, Aubrey K. Reid, Jr., made it even easier to agree to. The programme only asked for 30–40 minutes of time a month, after all, and it just might contribute to the overall effort.

Emotionally, it was a very difficult program for many senior managers to actually put into practice. The excuses were varied and usually not even articulated since, at first, there was no attempt to verify the completion of PEET visits. Reid, however, was visiting with the people listed on his PEET sheets each month and he was finding out that it was both emotionally and intellectually fulfilling to do so. So he instructed the QHV Director to find out how many other senior managers were being faithful to PEET. The answer was that only a small minority of senior managers were getting out to talk to both people on their PEET sheet in any one month. Fortunately, Reid had a solution.

At the next monthly meeting of the 24 senior managers, Reid announced that he was adding a wrinkle to the PEET programme. "From now on," he said, "when you finish your two PEET visits, fill in the bottom of the PEET sheet with any comments you have and return it to the director of the quality process. I've directed him to give me a report on the first of each month, letting me know who made their PEET visits and who didn't. Now, if you can't find time to make your PEET visits, that is your choice but do be prepared to explain to all of us here why you couldn't find 30 or 40 minutes all month to do this." PEET visits multiplied immediately.

And then the transformation began. After several months, the senior managers reported that they greatly enjoyed their PEET visits, that they were excited to find out what talented and interesting people there were on the payroll, that they had learned things about the business and about the competition that they had not previously suspected, and that they now better understood how the organization actually worked.

On the team leaders' side, similar reactions were reported. They were as surprised to find out that the senior managers were interesting, knowledgeable, and sometimes even funny as the senior managers were to learn the same was *true of them.* And they too reported having a better understanding of how things worked in the upper reaches of the company. Casual conversations across levels and across departmental boundaries became common. The senior managers knew their own company far better, and employee hesitation about bringing up an idea decreased because they believed that the senior management truly wanted to see improvements made in every aspect of what the company did. The emotional acceptance of the PEET programme followed the intellectual acceptance and made it possible for the company to make a major step toward its goal of being a quality company.

That is the nature of transformation. For permanent change in humans, there must be both emotional and rational acceptance and there must be a defined goal.

Did it work? Well yes. Within three years, the income from the company's primary product nearly doubled, while the staff increased by just four percent. After five years, the company was one of only two finalists for the Malcolm Baldrige United States National Quality Award in that award's inaugural year, in the service category. And, within eight years, the company's market share grew from 11.8 percent (second place) to 18.4 percent, with no other competitor having even half that much market share.

The Paul Revere Insurance Group did not, of course, invent this transformation sequence. Virtually every military service in the world employs the same technique during its indoctrination period for newly arrived members. The goal of such operations (sometimes called 'Boot Camp') is to instill values and technical and leadership skills through directed behaviour, resulting in some truly wondrous transformation at the individual level, once the recruit accepts the philosophy surrounding his or her new behavior.

The common thread between such apparently diverse examples as an insurance company's PEET program and a military organization's Boot Camp is leadership, a trait that knows no organizational boundaries, that can be found—or found to be lacking—in any group of humans.

Take as a definition of leadership the following: leadership is the creation of an environment in which others can self-actualize (see Masiow's hierarchy of needs) in the process of completing the task. Looked at in this way, the role of leadership in effecting organizational transformation becomes more obvious. Managers tend to be skilled at maintaining the *status quo*. Leaders, with their dual purpose of making it possible for others to self-actualize and to complete the task (which may well be changing as the transformation proceeds), engage others in their vision of the future and work to make it a shared vision. In short, they make it possible for others to become both intellectually and emotionally committed to the future of the organization, not simply the procedures and desires of the past.

One of the interesting aspects of the relationship between leadership and transformation was illustrated by Aubrey Reid in the Paul Revere case cited above. An important piece of his vision for the future of the Paul Revere Insurance Group was that participative and delegative leadership should be the primary leadership styles in use on a day-to-day basis, with the third option – authoritarian leadership – being used only rarely. This sort of relationship between people at all levels is necessary in order to establish a quality process that is to have any longevity. Secondly, he recognized that the company would need leaders at every level, and that if all decisions were being made at the top of the corporate structure, this would also run counter to his long-term vision.

Paradoxically, it took a series of authoritarian decisions such as the one described above to cause participative and delegative leadership styles to become the dominant leadership styles in the company. And, of course, it was the practice of these styles (by Reid himself and by those he directed) that set the example that worked to convince people, at every level of the organization, that they should begin acting as leaders as well.

When the senior people of an organization succeed in creating an environment in which their juniors can both self-actualize and accomplish their tasks, it encourages those juniors themselves to begin the difficult but ultimately satisfying work of becoming effective leaders. Within a relatively short period of time, the organization will begin to approach the ideal of having leaders at every level.

Doing the right thing (as defined by a senior individual and acknowledged intellectually) can precede emotional commitment. Once both aspects are fully engaged, transformation is under way. While perhaps not complex, it is difficult. There can no sooner be "transformation by proclamation" than there can be "quality by proclamation."

Transformation will not be achieved through the issuance of an edict, no matter how well-reasoned or provably necessary the desired change may be. An organization will need leaders capable of working with humans, who understand that humans are both rational and emotional and who are willing to meet them on both levels, and who are willing to do the hard work of leading—creating the environment, ensuring the proper training takes place, defining the goal and the tasks, being open to suggestions—and by saying 'thank you' when appropriate.

Transformation is the right word because the change is permanent. That is not to say that the organization will never change again—only that it won't be going back. As in chemistry class, more elements can always be used to enrich or alter the compound, but it is not possible to return to the basic elements the way they were before the transformation.

The International Journal of Business Transformation,
Vol. 1, No. 4, *Apr. 1998.*

FOLLOWERSHIP: AN ESSENTIAL ELEMENT OF LEADERSHIP

The phrase "too many chiefs and not enough Indians" needs no explanation, but in addition to its ethnic insensitivity, the statement also misses from a managerial standpoint. In organizations aspiring to growth and continual improvement, relationships are more complex and options more numerous.

Virtually no one is a leader all of the time. Quite simply, leaders also function as followers; everyone spends a portion of their day as a follower and another portion as a leader. For example, a senior vice-president may fill the role of "powerful person" when dealing with subordinates, but not when dealing with the president of the company. In turn, a president of a company who refuses to respond to others will soon earn the ire of his or her board or a legion of customers—and down either path lies despair.

It is more illuminating to think of the relationship between followership and leadership as two points on a continuum, anchored on one end by "passive followership" (a phrase from Robert Kelley's book, *The Power of Followership*—roughly equivalent to a slug) and on the other by "capital-L leadership" (a phrase from *Five-Star Leadership*—roughly equivalent to a prophet). In between, moving from passive followership to capital-L leadership, a person passes through "active followership" and "small-l leadership."

Most people spend their entire lives moving back and forth between these latter two. Small-l leaders and active followers have a great deal to contribute to those occasions when everyone "just pitches in and gets it done." It's called teamwork. At these times, the roles of leader and follower change so frequently that they're not worth labeling. When properly prepared, people who are not normally comfortable being labeled leaders will assume the role whenever their leadership skills are called for.

Unfortunately, most American businesses concentrate their training efforts (if they train for leadership) on capital-L leadership. It is, after all, the more exciting option. It is the sort of leadership exercised by women and men with the power to make decisions with far-reaching impact; it is the stuff that legends are made of. Small-l leadership, on the other hand, involves decisions that have immediate impact, usually on people known to the decision-maker.

Neither stands alone. Habitual capital-L leadership mistakes easily obliterate the bottom-line impact of a series of small-l leadership decisions; lackadaisical small-l leadership acts easily neuter solid capital-L leadership decisions. Any leadership curriculum that glosses over small-l leadership creates a void. So, too, does any curriculum that overlooks followership. If a person has not been trained—formally or informally—to fill the role of follower, the odds are significantly lower that he or she will ever reach full potential as a leader.

A look at an article written by Sergeant First Class Michael T. Woodward for the US Army's *Infantry* magazine in mid-1975 suggests the scope of a followership course. SFC Woodward points out that followers need to be committed to the mission of the organization, which in turn requires that followers know the mission and concur with its aims. This simple idea is, of course, a major stumbling block in organizations that demand blind obedience from lower-level employees. In fact, a decision to create an environment in which employees become active, committed followers requires real effort on all sides and more than a modicum of trust. SFC Woodward's goal is to create competent followers able to estimate the proper action required to contribute to mission performance and, in the absence of orders, to take that action.

SFC Woodward includes "Ten Guidelines for Followers." His starting point is the United States Army's leadership principles and the list reflects how close active followership is to small-l leadership:

1. Know yourself and seek self-improvement.
2. Be technically and tactically proficient.
3. Comply with orders and initiate appropriate actions in the absence of orders.
4. Develop a sense of responsibility and take responsibility for your actions.
5. Make sound and timely decisions and recommendations.
6. Set the example for others.
7. Be familiar with your leader and his job, and anticipate his requirements.
8. Keep your leaders informed.
9. Understand the task and ethically accomplish it.
10. Be a team member – but not a yes man.

In the words of SFC Woodward, "Effective leadership requires followers who are more than Pavlovian reactors to their leaders' influences. When followers actively contribute, are aware of their function, and take personal pride in the art of followership, then the joint purpose of leadership and followership—higher levels of mission accomplishment—is achieved effectively." He concludes, "Professionalism in followership is as important in the military service as professionalism in leadership."

Educating people to help them become productive followers and productive leaders is one of the responsibilities of leadership. Any thoughtful leader has three top priorities:

- Accomplish the mission
- Take care of your people
- Create more leaders.

From the senior management of the organization there must be a willingness not only to invest money and resources, but also a willingness to take part in discussions and set the example. They must come to grips with the continuum from followership to leadership rather than present the two as facing one another across a yawning gap. Otherwise, folks caught on the followership side of the chasm will show no ambition to "become leaders." The distance will be too great for all but the greatest leaps of faith.

Quality Digest (Internet), *Dec. 1997*

CREATING MORE CREATIVITY

Worcester County in Central Massachusetts may seem like an unlikely "cradle of creativity"—yet evidence exists to support the assertion of former Massachusetts Senator George Frisbie Hoar that, "Within a radius of 10 or 20 miles of Worcester sprang as great a group of creative geniuses as sprang from any group, anywhere in the world."

Historic names are among the more famous examples: Robert Goddard, inventor of the first liquid-fueled rocket, fired it successfully on his aunt's farm in the town of Auburn; the inventor of the cotton gin, Eli Whitney, born in Westboro; Luther Burbank, pioneer of fruit and vegetable research, born in Lancaster; Elias Howe, Jr., inventor of the sewing machine, born in Spencer. Other inventions of equal impact come without familiar names attached: the birth control pill, the use of ether as an anesthetic, a practical X-ray tube, mass production of barbed wire, the typewriter, the monkey wrench.

The earth doesn't have to shake But the creative process isn't just about BIG inventions; it is also about making life just a little bit more enjoyable or a little easier than it was yesterday. Along those lines, Worcester County has been home to the first marketing of the valentine card, a commercially practical envelope-folding machine (*followed by the first machine that also put glue on the flaps, driving the cost of producing envelopes from 60 cents per thousand to eight cents per thousand*), the square-bottomed paper bag (*including the machine to make more*), shredded wheat (*shredded wheat drink and shredded wheat baby food never caught on*), the rickshaw (*a local missionary wanted a simple mode of transportation to take to his posting in South America; from there it went to Asia*), the steam calliope, and factory-made piano wire. And no list would be complete without mentioning the yellow smile button and the curve ball.

A skunk-works or mountain top isn't necessary either Despite being a hot bed of creativity, Worcester County impresses a visitor as being an ordinary place. Which it is. Creativity does not require an esoteric environment, but it does require an appreciation of the possibilities and the time and materials to experiment.

But values that support creativity are necessary Once any community is established in which creativity and innovation are valued rather than viewed as weird or threatening, as long as there is time and physical support, anyone can get into the act. This is true whether the community is a county in Central Massachusetts or a factory in Idaho.

That leaves corporate America with a double problem. How can a company or an organization make creativity a priority? And what is necessary to support it? Quality can provide answers to both questions.

Establishing a Community that Values Creativity and Innovation

Granted, there is a certain amount of innovation and creativity that is going to happen in an organization with or without a quality process. Necessity (*if not desperation*) has long been accepted as the mother (*or, at least, the midwife*) of invention. There are also individuals who cannot leave well enough alone, who never take "no" for an answer. Some ideas, however, need a more forgiving, more encouraging, environment to emerge from the recesses of a potentially inventive mind.

The "community" that supports quality improvement supports creativity too... The pay-off of a quality process in regard to innovation and creativity is in its ability to establish the type of "community" that leads to a dramatic increase in the number of beneficial ideas that come to life within the corporate walls. A true quality process is founded on the premise that all of the individuals who make up the company are not only capable of thinking; they can be trusted to do so as well. That removes the barrier between the average employee and the creative process.

Make no mistake, this barrier still exists in too many organizations. The belief that creative ideas can come from places other than the executive suite and the research and development department is, all too frequently, considered to be a "daring" contention.

A quality process forces recognition that the intellectual capability to improve the organization is already there, resident on the payroll. It frees that intellectual power at all levels of the corporate ladder by insisting that everyone must be engaged emotionally with the proposition that continuous change and continuous improvement are part and parcel of who they are, of what they do every day. It also provides a focus for that change—quality.

Quality questions drive creativity A deeper understanding of quality is also an appreciation of the possibilities of improvement (*it's more than meeting specifications or controlling variation*).

There are at least two ways to look at every aspect of what happens inside an organization: "Are we doing the right things?" and "Are we doing things right?" Every one in every position of an organization faces this same duality. The impact of a decision that ignores either of these questions can be devastating at the executive level.

Focusing on doing things right... Focusing exclusively on doing things right could lead to making obsolete products perfectly. It may be a satisfying intellectual challenge, but it is also poor business.

Focusing on doing the right things... On the other hand, companies whose sole emphasis is on "doing right things" tend to zoom onto the scene with the newest technology and, after some heady times and flashy headlines, zoom right back out when someone else finds a way to do that same thing more effectively. Being a forerunner does not guarantee winning the race.

The point of asking "Am I doing the right things right?"... Less devastating, but equally counter-productive, are routine operations within an organization that fail to take both questions into account. A clerk may fill out 1,000 forms correctly, but if only 30 percent of the information on the form is useful, 70 percent of the effort is wasted. In such a case, it is likely that the clerk has some inkling that something is amiss. If trained to ask both questions, the clerk's inkling can be converted to certainty—the first step toward the possibility of change.

The case of Pontiac 6000... Another scenario pits the executive suite against the factory. For example, the Pontiac 6000 is the only American-made car model to break into the top 10 in the 1991 J.D. Power ratings of cars based on complaints during the first few months of ownership.

The model was first introduced to the American public in 1982. In the following years, while top management apparently failed to come to grips with the question of "Is continuing to make the Pontiac 6000 the right thing to do?" the people for whom making the model was defined as the "right thing" worked continuously at making it right. Finally, eight years after its introduction, it joined the list of best made cars in the world, undoubtedly due to a whole series of improvements and innovations between 1982 and 1990.

Unfortunately, between the time the data to determine the standings was gathered by the J.D. Power surveyors and the publication of the list, production of the Pontiac 6000 was cancelled.

Just when the employees of Pontiac were finally doing it right, top management decided it was no longer the right thing to do.

Battering Down the Barriers isn't Enough

Even in organizations where the corporate culture has removed conceptual barriers and provided a clear vision of what they are trying to accomplish, more is needed.

Day-to-day support in the form of training and structure is necessary to convert possibilities into realities. Add to this the empowerment, recognition, and communication processes that are considered part of a typical quality process and you'll find they also foster creativity, either as a goal or a tool.

Quality training... Take a look at quality training. In most organizations, employees are taught about quality itself (*awareness*), followed by problem solving techniques (*creativity*), and ways in which the company itself is structured so that they can contribute (*action*).

Quality structure... In companies where the team approach to quality is used, how to conduct a meeting in a participatory manner is part of the curriculum. If a suggestion box is used, how to write up a suggestion is standard fare. In either case, the goal of the training is to enable people to accomplish continuous improvement.

Support when they come on board... Or look at orientation classes for new employees. Unless taught not to (*at their new company or, perhaps, by previous employers*), new employees have the delightful tendency to ask questions such as "Why do you do it that way?" and "Wouldn't it be possible to do it this way?"

Primed to ask these questions, interacting with employees who are aware that "just because" is not answer enough, those questions can be the seeds of improvement and innovation. But remember that they need to be acted on as quickly as possible—before a new employee drifts into the habits of the "way we do things around here."

Training the middles Often neglected, but of vital importance, is training for middle managers. Middle managers are the primary keepers of the corporate memory. They need to be trained to appreciate the value of what they know, while at the same time, realizing that they need not be an authority on everything. They can be both a source of creativity themselves and a marvelous sounding board and champion for those who have the beginnings of an idea.

Overlooking or under-estimating the middles can be devastating... This is a new role for most corporate managers, and one many look on with skepticism. A wise company courts middle managers, both for offensive and defensive reasons.

On the one hand, if creativity and innovation are going to take hold and become habitual, middle managers will have to be active proponents. And, for that to happen, they will have to understand their new role and what is in it for them.

Top management needs to spend time with the middles to create the new "community"... Reaching that level of understanding will not happen accidentally. There will need to be sessions, both formal and informal, conducted by senior managers to define the new role of middle managers as coaches, mentors, researchers, facilitators, and cheerleaders. Their essential contribution in creating the new environment must not only be acknowledged, it must be worked out in a series of participative discussions.

Too often, middle management is not convinced that improvements are either desirable or possible. They can bring a quality effort to a halt by simply ignoring it, or by adding their own "don't get suckered— nothing has really changed" spin to all communications coming down the corporate ladder and an equally insidious "these people haven't got any ideas—nothing has changed" spin to all communications going up.

Technical training is a key to innovation too... Technical training can also trigger creative innovation. The confidence that comes from doing a job well can be translated into the confidence to look at how to do the job better (*competence is the first step toward excellence*).

Courses specifically aimed at awakening creativity can also be included under the umbrella of the "quality curriculum." In short, all the investment in training inherent in any successful quality process also supports the desired increase in creativity and innovation.

SENIOR MANAGEMENT'S ROLES IN EMPOWERMENT AND CREATIVITY

Once employees are prepared to participate creatively, they must be willing to do so. That is the essence of empowerment, and it depends as much on emotion as intellect. Again, senior managers must set the tone.

Driving out fear... One of the major contributions that top managers can make to the establishment of a creativity-friendly environment is helping people at all levels of the company understand that to be creative will not be seen as a threat by those senior to them.

When employees believe that it is safe to become emotionally involved in quality improvement, one more barricade to creativity—fear of being punished for "unauthorized thinking"—will be eliminated.

Communicating and rewarding quality and creativity... Even if they don't happen to be particularly creative themselves, senior managers can also foster, publicize, and reward creativity wherever they find it in their company. This will be far more likely to happen in the context of a quality process than in typical, top-down bureaucracy.

Saying "thank you" can be powerful... Most recognition schemes in a quality process are based on saying thank you to employees for specific changes they have made. This recognition need not be in the form of bonuses and pay raises. Names in a company publication; a visit from a senior vice president; free sundaes; parity in the company cafeteria: all can demonstrate to lower-level employees that innovation is an activity that is valued.

Celebration of innovation is a powerful inducement to continue using creative talents, dramatically increasing the odds that everyone will begin to allow themselves to think while on the job.

Forums for information sharing... Communication is vital to creativity. Information keeps the juices flowing, no matter which way the interaction occurs: up, down, lateral, it all contributes.

Even in Worcester County, the importance of having a sounding board for ideas is recognized. The Worcester County Inventors Club meets monthly. Described as "a pretty motley crew" by Edith Morgan, editor of their newsletter, the members include students, professionals from area R&D departments and people who just tinker—drawn together to share experiences and discuss common problems.

Corporate America is learning to appreciate what a team effort can do. According to *Management Practices: US Companies Improve Performance Through Quality Efforts*, a GAO report, a growing number of companies are moving away from individual suggestion systems to teams.

Case example: low-tech driver for high-tech innovation... The medium of communication can itself be an outlet for creativity.

A low-tech flip chart hanging on the door of the team leader's office is the creativity cornerstone for one of the most successful quality teams at the Paul Revere Insurance Group (*a team of computer technicians who, in one year, implemented over 165 ideas and saved the company over $360,000*).

Any time, day or night, that anyone had an idea—about anything, no matter how incomplete the thought—they would immediately write it on the flipchart. The team leader (*who was also their supervisor/middle manager*) then finds someone who will be willing to investigate and, possibly, implement the idea.

In the course of the year, the team held two formal "sit down with an agenda" meetings; the team leader held hundreds of two–and three–person meetings. Unorthodox, but effective.

The middle manager's role in increasing information about quality and creativity... In a quality process, middle managers find communications to be a substantial portion of their new role. One way in which middle managers can encourage the continuous growth of creativity in their subordinates is by frequently exposing them to new ideas.

These ideas can be garnered from professional reading, picked up at off-site conferences and meetings, or "stolen" from other quality teams in the company.

In the latter case, such thievery should not only be encouraged, it should be facilitated by the quality department. Make a list of current quality team ideas/projects and publish it as often as possible.

Anything that can be done to insure that people feel that they are part of a creative community should be done.

Communications is one of the trademarks of a successful quality process for a very simple reason: it works.

The expression "managing by walking around" has quickly become a cliche precisely because it is so effective. Not only do top managers get a better look at what the non-management employees actually do day-to-day; lower ranking members of the organization get a better look at what is going on in the upper reaches of the company. This information sharing becomes the basis for better decision making at all levels.

The basic building blocks of quality and the basic building block of creativity are the same. In creativity as in quality, an innovator must have a clear vision, technical capability, courage and drive. Goals, training, empowerment and recognition can draw a company together with the realization that each and every member of the organization can build a better mousetrap.

Journal for Quality and Participation, *Sept. 1991.*

13

KEEPING QUALITY ALIVE

As with virtually any living organism, a quality process needs attention, needs "care and feeding," if it is to survive to flourish. There never comes a point when it is safe to assume that quality will continue to be the source of benefits even if it is ignored or taken for granted. The senior management (including those people who have personal responsibility for overseeing the day-to-day operation of the quality process) must remain vigilant, must keep learning new concepts and procedures, and must introduce appropriate ones throughout the organization.

A quick test of the health of your quality process can be conducted by simply answering the five questions posed in "Are You Practicing Total Quality? Take the Test." One method might be to send copies of the questions—just the questions—to members of your senior management team prior to a meeting and make the discussion of their answers the sole agenda item for a subsequent meeting.

There were signs as early as 1991 that the pursuit of quality was slipping down the collective list of corporate priorities in too many organizations. Some of the culprits in this decline are targeted in "Warning: This Good Idea May Become a Fad."

The two-article set titled "Remaking a Quality Management System, Part One" and "Remaking a Quality Management System, Part Two" serve both as a succinct summary of the perils of lassitude and as a guide for breathing life back into a process that has slowed considerably, if not come to a complete stop.

ISO 9000 and other efforts initiated in the name of "quality" have caused executives and organizations to lose their focus. The four negative forces that have had the most to do with delaying the worldwide quality movement are detailed in "What Happened to Quality?."

Quality can be introduced or re-introduced into any organization so long as the senior managers can convince their employees that the effort is serious, that their commitment is real and will be sustained, and that the long-term good of the organization and the individuals who make up that organization depend on the success of the effort. If nothing else, previous attempts at quality can simply be referred to as learning experiences or pilot programs while declaring that, "This time quality will be *done correctly*."

ARE YOU PRACTICING TOTAL QUALITY? TAKE THE TEST

Like most management feeding frenzies, Total Quality Management hasn't made for a very satisfying repast. Most companies follow the recipe religiously, but lose the flavor and aroma of true quality.

Want your outfit to cook up a winner? We believe fewer than 10 percent of U.S. companies can successfully pass this test.

1. Looking at the last 10 decisions made by senior management in the name of "total quality," whose beha- vior was expected to change?

   If it was always, or usually, someone other than the senior decision makers, then they're not practicing total quality, they're practicing quality by proclamation.

2. In your organization, is there an easy answer to the question "Where do I go with an idea for quality improvement?"

   It is remarkable how few organizations have clearly defined avenues for employee participation. In a recent survey of 2,800 federal installations, more than half said they were practicing TQM. Yet the average number of employees actually doing anything to help the organization improve was only 13 percent.

   If fewer than 100 percent of the people on the payroll are literally, formally, structurally, continuously enrolled in the quality process, forget about "total" quality.

3. Are small ideas captured?

   How long does it take to implement an idea for shaving, say 15 minutes a week from your operations? If it takes significantly longer to get an audience for the idea than the idea is worth, you're in trouble. Any process geared to look only at big ideas, taking months of careful study, isn't up to snuff.

4. When is the last time someone in the top layer of management personally said thank you to someone in non-management?

   If it's more than a week, you fail. Sounds pretty basic, but it's more than a measure of good manners. Simple recognition shows interest in the people who deliver quality—and a degree of personal interaction essential to the quality process.

5. When was the last time a member of top management read something longer than three paragraphs about quality and continuous improvement?

   If the answer is longer than a month, you're slipping backward. The best measure of an abiding commitment is continuing investment of time and thought.

If you can't answer all five questions positively, your quality effort is less than total; that is, you flunk!

So what's the payoff if you pass the test? For one, avoiding problems like the following: When the Saturn became a commercial success in 1991, management showed themselves to be true GMer's at heart by attempting to crank up the production numbers to meet burgeoning demand.

The workers at Saturn's Spring Hill, Tennessee, plant threatened to strike. Why? Because they were trying to maintain their hard-won gains in quality—and were too proud of what they'd accomplished to let volume-driven mandates degrade their product. Management backed down, and production numbers are now climbing. Better yet, customers who choose Saturn for its value still get it.

On Achieving Excellence, *Dec. 1992*

WARNING: THIS GOOD IDEA MAY BECOME A FAD

There is no doubt that quality is "in." Since World War II, a growing cadre of devotees has labored to bring quality to the attention and acceptance of corporate America. Today, interest in the concept of quality and recognition of the potential results of an organization-wide quality effort are at an all-time high...and rising.

Paradoxically, to use the all-but-forgotten phrasing of the teenagers of the beloved 50s, quality is so far in, it is almost out.

Euphoria is becoming tinged with unease. There is ample cause for both. Taking a minute to celebrate (always appropriate for quality), there is great satisfaction that the basic tenet of quality has penetrated the national consciousness: when organizations benefit, individuals benefit—both as employees and consumers. Yet, the very success of this message has obscured some fundamental truths:

- Quality is hard work
- Quality must incorporate every member of an organization in decision-making to reach its full potential
- There is no one-size-fits-all approach to quality.

Unless these issues are faced directly in the next decade, quality will be perceived as over-promised and under-delivered...just another business fad.

HOW BAD IS THE SITUATION?

Quality is still on the upswing. Fueled most recently by the advertisements of Baldrige winners Cadillac and Federal Express, awareness of quality now pervades all sorts of organizations previously self-declared to be too "different" to be able to benefit from a quality process. Education, health care, government are segments of society just beginning to explore ways to bring the gains associated with quality to their own organizations. But the seeds of disillusion also have been sown.

TQMism Take the example of total quality management. It is subject to blatant abuse, the most flagrant of which is to reduce it to sloganeering; a pitfall against which Dr. Deming has long cautioned.

Who hasn't heard the acronym TQM, as in "We're really into TQM at my outfit," reduced to business-crowd, cocktail-1gathering chatter? It is almost a required phrase in order to hold one's own. Extra points are scored if you can make offhand mention of training you have been through (add points if you had to travel somewhere for the training). Bonus points are awarded those who can name their new vice president for quality and make reference to a conversation they had recently with her (or him).

Such chit chat is not inherently harmful, but it is all too often symbolic of a misapplication of TQM itself. TQM is supposed to stand for TOTAL QUALITY MANAGEMENT; all too frequently a more honest interpretation would be TIMID QUALITY METHODS.

Six breakthrough teams involved in diagnosing and repairing faulty processes is not TQM. If the phrase is to be used with integrity, it must be used literally. Total ...means everybody. It doesn't mean some of the people this month and some others next month or next year. It doesn't mean that everyone will eventually be impacted by the decisions of a few. It means everybody is involved in formulating decisions. Formally. Actively. Everyday.

Such involvement is virtually impossible to find. If it were possible to design a litmus test for a true total quality process, the process would have to meet at least the following criteria:

- Is every person on the payroll literally, formally, enrolled in the process? If not, the process in question should be labeled "beginner." It is a partial effort at best and will produce, at best, partial results.

The argument that everyone is impacted by the decision of a handful of teams is fallacious. Such an approach addresses only half of the problem; the other half is how to tap the talents of everyone on the payroll.

- Do decisions made as part of the quality process normally call for changes in the behavior of the decision makers or are the decision makers simply dictating changes for others in the name of quality?

If decisions always call for changes in the behavior of others (subordinates, of course), then management has simply found a new name for business as usual.

- Are quality efforts limited to a few "breakthrough" projects while the detailed and difficult assessments of whether the organization is doing the right things and doing them correctly are left to chance? If so, this represents another version of crisis control, not quality management.

Are late entries different? Higher education is a johnny-come-lately to the field of quality, and as a result, it has many examples from industry to emulate. The results are not encouraging. A comparative study by L. Edwin Coate of Oregon State University titled "Implementing Total Quality Management in a University Setting," (see December, 1990 *Journal for Quality and Participation*), gives data on 25 institutions of higher learning. None employ 100 percent involvement. Colorado State University had 50 TQM teams operating at the time of the July 1990 study; no one else even came close.

Half-loaf TQM Total quality is not for the faint of heart. It is self-delusion to implement only one aspect of quality and expect quality results. If too many half efforts are paraded about as examples of "what quality is," then quality will come to be seen as a fad, something to be endured until it expires of its own weight, followed by a management idea hiatus (when some real work can get done) followed, inevitably, by another fad.

Awardism Another grave danger to the overall quality movement in America derives, unfortunately, from the very thing that has been such a major contributor to making quality a household word: the Malcolm Baldrige National Quality Award. Specifically, the danger is being introduced by the growing band of "Baldrige consultants."

Baldrige strengths... One of the great strengths of the Baldrige is that it is not prescriptive in nature. Consider for a moment the Deming Prize. It has been in existence for almost 40 years and yet no one has ever seriously tried to duplicate it. Why? Because it is highly prescriptive—an approach which obviously has worked well in Japan; for there is no questioning the Deming Prize's key role in Japan's quality revolution.

Yet in the less than four years of the existence of the Baldrige, there are clones of the award in hundreds of companies and associations, dozens of states, and a rapidly growing number of countries.

Why? Because it is descriptive of desired results, rewarding creativity and diversity, leaving individual organizations great latitude in determining exactly how they will achieve quality. The application is less a one-size-fits-all straitjacket than it is a quality audit, a compilation of the questions that require answers if an organization is going to produce quality products and service.

*Beware the "expert"...*Baldrige consultants will endanger this freedom of choice if they coalesce into a *de facto* JUSE (the Japanese Union of Scientists and Engineers, which controls the Deming Prize) and the perception builds up that there is one "correct" answer to every question—or that it would be impossible to win a Baldrige without a consultant's knowledge and backing.

Any consultant who claims to have the inside track, who claims to be able to all but guarantee a site visit, is, of course, blowing some serious smoke. The corps of quality professionals knows this, but the legions of potential customers for such consultants don't know it. Failure to win under these conditions can be serious breeding ground for cynicism.

There is no easy solution. If winning a Baldrige becomes more important than digesting and using the results of the quality audit for further improvement, the Baldrige itself will be greatly devalued. One can only hope that 'wannabe' Baldrige winners in the next few years will have the self-confidence to define their own quality processes based on individual judgment combined with well known quality principles—and avoid self-proclaimed Baldrige experts.

QUALITY IS AT A CROSSROADS

Its very success has raised expectations, yet niggling doubts are beginning to surface. Quality professionals have a role in informing customer expectations, even when that role is unpalatable. No one can do quality for an organization or to an organization, only with an organization.

Nothing less than a total quality process will suffice to produce the desired results. These are messages that no company, hospital, university or bureaucracy wants to hear. But, unless the message is clear, skepticism about quality will continue to grow.

The challenge for the future is to consolidate the gains quality has made in the past, to capitalize on the hard-won recognition of the benefits of quality, and to refuse to allow indifferently applied quality methodology to usurp the field. An even more critical task is to look for opportunities for quality to evolve. It would be a dreadful irony if a discipline based on continuous improvement were itself to become so hidebound that it rejected new ideas.

Journal for Quality and Participation, *Mar. 1991.*

REMAKING A QUALITY MANAGEMENT SYSTEM, PART ONE

What if you already have a quality process in place, albeit with results nowhere near what the consultant promised all those months—or years—ago? Has your process taken on a comfortable, low-level life of its own with no apparent hope for improvement? Then what good is advice? How can you possibly overcome the negatives?

This column, the first of a two-part set, addresses those questions. We'll focus this month on rethinking efforts so change can be considered. Next month will examine the mechanics necessary to improve a struggling quality process. Both columns will offer ideas on what makes a successful quality pilot, one that prepares a company to operate on a more expansive, but less expensive, scale. Only then can something new and better be put in place.

First, though, through informal conversations and formal presentations, do a companywide reality check. An organization's top management must maintain a unified vision of necessary and possible quality measures. This common wisdom includes the following basic points.

- *Quality is both rational and emotional.* Rational efforts stress measurement, while emotional efforts concentrate on rallies and hoopla. A quality process needs a balance of both for a very simple reason: Quality processes depend on people, who are made up of both rational and emotional elements.
- *Quality won't go away.* Despite all the disclaimers and naysaying, quality won't go away because good customers won't let it. In fact, they will punish any organization that tries to pull back by taking their business elsewhere and spreading the word. All of us helped create the quality revolution, and none of us can expect our customers to be indifferent to it.

- *The four essential reasons for "doing quality" drive every successful enterprise:* Quality makes money, makes customers happy, makes employees happy and makes sense ethically.
- *The concept of quality is not complex.* Implementing quality requires constant attention to details, but the average person can easily grasp the concept.
- *Service to external customers rarely exceeds service to internal customers.* Concentrating quality efforts solely on front-line people will prove counterproductive. If employees aren't treated respectfully by an organization—if they don't receive quality service themselves—they won't treat their customers consistently well, either. That's true no matter how many smile classes they've sat through.
- *A quality effort should ask, "Who can we afford to exclude from our improvement?"* The answer is, of course, no one. Unfortunately, most quality efforts ask instead, "Who should we include in the effort?" Down that path lies underachievement on a grand scale. Only 100 percent employee involvement offers the greatest success.
- *Companies must consistently pursue quality.* Simply announcing. "We are a quality company" and making that part of an advertising compaign won't create a quality organization. Quality by proclamation doesn't work.
- *Executives must realize that their personnel departments hire adults—and they deserve to be trusted.* Unless and until individuals demonstrate they aren't worthy of trust, they should continue to take responsibility for their own work.
- *Each organization's quality effort differs in its details.* A one-size-fits-all process doesn't exist, even though quality principles can be universally adopted. A company must adapt specific techniques, incorporating the best and most appropriate principles from

other's efforts. A 100 percent employee participation level should be adopted upfront; exact techniques will vary. That's why hiring an outside consultant on a long-term basis invariably wastes money.

- *It shouldn't take forever to see results.* The elapsed time between concluding a two-day quality workshop for senior executives and implementing an effective quality management system with 100 percent employee involvement should take six to eight months for an organization of 3,000 employees or fewer.

Quality Digest (Internet), *May 1998*

REMAKING A QUALITY MANAGEMENT SYSTEM, PART TWO

Last month, we began a two-part description of what an organization might do to snap out of a "quality slump"—a period when a quality process is in place but with little to show for it. This month, we'll look at the essential ingredients for a successful process.

First, three questions must be answered regarding quality: Whose behavior changes? Where do I go with a good idea? And how will taking this measurement help me? In many organizations, the answer to the first question is, "Everyone except senior management." Those same organizations believe there is no real answer to the second question, and merely asking the third question is considered incredibly naive.

What's the best answer to the first question? Everybody's behavior must change, starting with senior management. In stalled processes, odds are that a quality consulting firm implemented the effort. The consultants put a flock of senior managers through training and then excused them from participation, barring an occasional speech. Meanwhile, the consultants went on to work directly with the middle managers and below to implement the systems. Much expensive training may have followed.

It's time to put training dollars to better use. Nothing significant can happen until an organization's senior management gets personally involved—and stays involved. Senior managers must come to a collective agreement about what quality is and whether they are ready to give it the necessary support. This is one of the few points in the process when a company can financially justify importing one or two "quality experts." However, outsiders must function more as facilitators and advisors than as consultants giving prepackaged answers.

Typically, these experts will conduct a two- to three-day workshop with the senior management team to help them organize their thoughts, increase their knowledge about quality and present them with optional methods of defining and implementing a more complete quality process. They will help them understand the potential benefits of a quality process and reach consensus on a company vision in which they are personally invested.

Executive training must be followed by executive action. If there is no answer to the second question—Where do I go with a good idea?—it's time to decide on one. The best way to deal with this is to establish a committee composed of senior managers who can make decisions about allocating resources, both people and funding. More often, middle managers are delegated the task of defining an organization's quality effort and told to take their recommendations to senior management.

The message sent was—and still is—all wrong. Those decisions must result from in-depth discussions and conscious decisions by the senior managers themselves. If there is no desire for a fundamental change in the way the organization does business, then the whole effort should be abandoned.

For as long as it takes, the management committee should meet weekly in the organization's most obvious meeting place. The rest of the organization must see that the folks at the top think a quality process is so important and that they will devote two to four hours every week to it. If attendance lags, the entire company will know within hours. Our advice? Make attendance mandatory, and cancel the meetings only if the building is on fire.

The committee has a number of choices: They can train current management to proactively solicit ideas, they can establish an effective suggestion system, or they can implement a team approach. When the second question is asked, every person in the company should know and believe the same answer.

Any answer on where to go with a good idea will require support. That includes:

- *Someone to track results.* A few hardy souls can do the job. One of the most successful quality teams in US corporate history consisted of a manager, four analysts and a part-time secretary. They managed by having, as a fundamental tool, a computer-based method to track quality improvement ideas, their current status and their impact.
- *Appropriate training.* If a suggestion system is used, training must include such basics as how to write a suggestion, how to solve problems by using appropriate measuring tools and understanding quality fundamentals. If teams are used, training must include how to run a meeting in a participative manner and basic leadership skills. In fact, no matter what the approach, a program for improving leadership skills throughout the organization should be implemented. Find a training company with a product that can be customized and then, after the course has been conducted a few times, buy the rights to the program and integrate it into the organization's regular training cycle. That way, next year's attendees won't simply get the same course as this year's; they'll get a better one reflecting the first year's successes and lessons.
- *Effective use of measurement.* This leads to the third critical quality question: How does taking this measurement help me? The answer is built into quality's fundamental role: to create competitive business. At the most basic level, what's in it for you when you use measurement to keep a company in business is a job—your job. If that connection hasn't been made, it's time to make it.

On the other hand, if an organization has previously used measurement as a weapon or religion, it's going to be difficult to make that connection. When employees are asked to take measurements that supply answers to questions that are never asked, they are bound to be skeptical. Measurement has only two legitimate uses: as a source of ideas and as a way to track progress. Involving employees in designing measurement can help ensure that measurements are used effectively.

Linking measurement to recognition also is a powerful connection, although not all recognition should be linked to objective measures. Organizations that recognize quality improvement of any kind encourage more of the same; organizations that ignore employees' quality efforts kill initiative.

So now you have the three questions that need answering. When senior managers have the answers, they should be announced in some memorable way. At a ceremony, the previous quality efforts could be acknowledged as a "successful pilot," and participants can be praised for proving the worth of paying attention to quality and helping to convince senior management that it was time to involve everyone.

The next step after the process is in place? Change. Any process must change, even a wildly successful one. After all, if a process that advocates change is not itself open to growth and improvement, everyone will quickly spot the inherent hypocrisy. One of senior management's primary tasks is to watch for appropriate times to introduce changes.

It is never too late to begin or fine-tune a complete quality process. Quality must not be allowed to become the privately owned property of a few nearsighted devotees whose vision does not go beyond preserving their own jobs. Quality is not something extra that is done by a select few when there is time. Quality is how work is done, and how relationships are formed and strengthened. It can only happen if everybody is involved and if the organization not only gives permission but encouragement and structure to the process.

The fact that the effort has started badly or has gone astray doesn't mean the organization is doomed to forever mourn what could have been. Quality is about change. A non-productive quality process can be a wonderful beginning.

Quality Digest (Internet), *June 1998*

WHAT HAPPENED TO QUALITY?

During the late 1980s, it looked as if quality would be the force unifying management and non-management. It would cut the Gordian Knot of industrial relationship and make it possible for companies to earn money, keep their customers and employees happy, and be ethical—all at once. It almost worked.

In a few instances, quality delivered for a while. One example: The Paul Revere Insurance Group in Worcester, Massachusetts, defined and implemented a quality process that hit all four measures for more than six years. Froedtert Memorial Lutheran Hospital in Milwaukee adapted Paul Revere's approach and continues to enjoy the results. Baldrige winners also prosper, but the number of organizations that have managed to meet all four criteria still doesn't reach triple digits.

For most, quality proved to be just another flavor of the month. It's not uncommon now to hear a US senior business executive say, "Quality? Yeah, I think we did that in '91. Or was it '92? Hired one of the leading firms to do it for us."

What happened? And—far more important—is it too late to save the quality movement?

First, as to what happened, four negative forces come into play: lack of leadership, bad beginnings, re-engineering and ISO 9000.

Perhaps the most damaging force is the presence of managers rather than leaders in the senior ranks of US business. Managers trust numbers; leaders trust people, who then understand and use numbers. The United States' B-schools have made a profitable science out of producing technicians who have risen to positions of power while having little, if any, association with non-management folks. Asking these executives to trust a degreeless junior employee with a decision that might impact the bottom line is akin to asking them to trust a Martian. How are they going to trust someone with whom they've never interacted? Leadership requires people skills, and leadership and quality are inseparable.

Bad beginnings also grow from emphasizing numbers. While it's true that a journey of a thousand miles starts with but a single step, if that first step lands in the wrong direction, then the trip will be a long and frustrating one. The American quality journey began with a combination of quality circles, W. Edwards Deming and a few large consulting firms. Quality circles aim too low by preaching the enlistment of 10 percent to 15 percent of the work force (What's wrong with the other 85 percent to 90 percent? Don't they have anything to contribute?) and ignoring active involvement of senior management.

Deming's magnificent humanity and uplifting emphasis on the individual worker's potential is ignored by virtually all of his "disciples" in their rush to define new measures and charts. Deming's statistical mastery is taken to define who he was. In fact, he was much more. Equally damaging, big consulting firms promise senior managers that, once they've attended an expensive course or two, they can get on with their own, more important work while the consulting firms define and run the organization's quality program.

Reengineering also makes a negative contribution to quality's reputation. Few people notice that reengineering is only a subset of a subset—specifically, a subset of measurement, which is a subset of quality. While reengineering is a powerful and necessary subset with the potential for a tremendously positive impact on the bottom line, it's not the final solution. Reengineering appeals to senior managers because it reinforces their belief that the only people capable of making meaningful decisions are senior managers.

Unfortunately, too few organizations realize all the gains reengineering promises. By reducing everything to process control and by laying off droves of people considered superfluous, many post-reengineering organizations find themselves with diminished capacity for growth and an extremely hostile work force. Reengineering's chief proponent, Michael Hammer, established new standards for understatement when he offered the assessment, "We forgot about the people." Unfortunately, disillusioned companies are quick to couple reengineering and quality, and write them both off as a bad job.

It's equally devastating when companies substitute ISO 9000 for quality. ISO 9000 is a lazy option. One well-documented paper estimates that ISO 9000 covers about 10 percent of what the Baldrige covers. However, ISO 9000's popularity is easy to explain.

In the '60s, '70s and early '80s, a subculture of "quality guys" who could manipulate numbers grew up in US business. Then came the call to involve people at all levels and occupations. As quality's definition expanded to include both the emotional and rational, it began to slip away from the traditional quality control community.

When ISO 9000 (and the rest of the rapidly growing ISO family) came along, the quality control types felt vindicated. Charts and checksheets and micro-controls were back in style; trust and leadership were no longer of prime importance. Unfortunately, ISO 9000 and quality aren't the same. ISO 9000 allows a company to make concrete life preservers using slave labor, as long as the documentation is adequate. Quality has a broader vision. By disenfranchising the majority of employees, ISO 9000 is a giant step backward.

Is it too late to save the quality movement? While the United States has squandered a wonderful opportunity to jump decisively ahead of the international business community using quality as a differentiation, quality's potential still remains. The

Baldrige continues to be a valuable tool, especially when considering the changes made in the application process during the last two years. US business's current interest in leadership also offers hope. Companies that encourage leadership at every level are halfway to quality. The measures listed earlier—making money, keeping customers and employees happy, and being ethical—are the measures of any successful quality effort. And there is no reason that all four cannot be achieved simultaneously.

Quality Digest (Internet), *Feb. 1998*

14

PUTTING THE FOCUS IN THE RIGHT PLACES

A reminder. There are four primary reasons for doing the hard work of quality: Quality makes money; quality attracts and retains customers; quality makes employees happy; and quality is the ethical thing to do. All of which only occurs if the process is well-defined, intelligently initiated, and enthusiastically implemented and maintained.

But the question still comes up: "What's Next After Quality?" In truth, the answer is: More quality. Once customers are used to receiving products and services that live up to their expectations, a competitor will offer products and services that exceed those expectations. In a remarkably short period of time, what was exceptional will become the norm...until yet another company makes an improvement or two and offers products and services that generate even higher expectations. The cycle is endless. Quality is not going to peak, much less go away.

There is still one question left to answer: "What's In It for Me?" It's a question that you can expect every employee at every level to ask, and it is a perfectly valid question. Fortunately, when a quality process is *done correctly*, there are equally valid—and attractive—answers.

WHAT'S NEXT AFTER QUALITY?

Total Quality Management has become a meaningless phrase. While some productivity-cum-quality consulting firms accept big bucks for churning out warmed-over quality circle programs, others reduce quality to a seven-step, cookbook approach to problem solving. Both leave organizations with a "been-there-done-that" hang-over and heightened employee cynicism.

In the absence of significant progress, managers ask a common but empty question: Now that we've "done quality," what next?

Quality done right.

Rather than farming out the problem to consultants, managers need to understand how and why quality works. Quality is simply common sense formalized into everyday procedures. And while each successful quality process is in its details, unique, all share three imperatives: leadership, participation and measurement.

LEADERSHIP

Lack of visible top management commitment is the root of employee cynicism about quality. True commitment goes beyond: "It looks like this will make us a lot of money; I'd better have my employees do it." It demands that senior managers invest their own ego, time, effort, and resources in improving whatever it is they do.

Building an integrated quality environment requires leadership at every level—not the bold, capital "L" Leadership addressed in most books, but rather the small "l" leadership concerned with day-to-day decision making and relationships. One little-used leadership-development blueprint comes from a profession that has studied leadership (both big "L" and little "l") for over 2,500 years: the military. Military leadership theory is an incredibly rich, often surprising resource. For example, the notion that leadership is a subset of love is perfectly consistent with mainstream military thinking on the topic. For a good introduction to the genre, consult the US Army's *Military Leadership* (see Recommended Resources).

PARTICIPATION

The most expensive cost of poor quality may be the wasted talent within an organization. This waste occurs in two ways: Excluding people is the most obvious. Conventional wisdom would have management ask, "Who should we make responsible for quality?" or, "Who can we get to volunteer for this quality thing?" This approach typically yields 10 to 25 percent participation. (All of the 1993 and 1994 Baldrige winners achieved 100 percent employee involvement.) Ask instead, "Who can we afford to exclude from our effort to improve?" Clearly, nobody. No quality scheme works unless it involves everyone; that means sharing power and information with all employees.

The second cause of waste is less obvious: You must engage *all the talents* of each individual to take full advantage of your resources. To that end, organizations must invent structures that allow everyone to collaborate in problem solving. Many create team-based organizations. Others follow Milliken & Co.'s example, using what is essentially a suggestion system. (Milliken received 67 ideas per employee in 1994 and implemented better than 85 percent). Any option that enables an employee with a good idea to receive a quick, simple response—followed by recognition—will suffice.

Because good decisions depend on good information, you must make a conscious effort to ensure that (a) more than enough information is communicated to all hands and (b) what is heard is what was intended. Anyone tempted to give communication short shrift because,

"We're small so I know our communication is OK," should consider all the couples who divorce due to "an inability to communicate."

MEASUREMENT

Old-line quality-control veterans threatened by the new quality revolution cling to measurement like a lifeline. In contrast, employees all too often see measurement as a weapon to be used against them ("Aha! Caught you!") or a religious ritual. ("We can record them, so we should; the one who dies with the most charts wins.").

For quality improvement efforts, measurements should be taken either to identify data that can be a source for ideas or to track progress. The key to successful measurement is to ensure that everyone understands why measurements are taken. It serves no purpose to make measurement mysterious.

WHAT NEXT?

Most quality efforts fail when organizations focus solely on leadership or participation or measurement. A fragmented approach is analogous to juggling one ball at a time. All the pieces must be moving at once for anything noteworthy to occur. To assess the likely effectiveness of any quality process, ask yourself three questions:

- Whose behavior must change? (Hint: Think senior executives.) If the chiefs don't change their ways, why should anyone else? The quality process requires behavior changes in the executive suite if there is to be any change in the mail room.
- Where do I go with a good idea? Simple as it sounds, this question is the key to active participation. Everyone in the organization must answer the same way.

The consistent, responsive handling of frontline input allows Toyota to implement 95 percent of employee suggestions.

- How will the numbers be used? Every measurement must be used either as a source of ideas or to track progress. Avoid data for data's sake.

Before hiring outside help, ferret out approaches that limit or complicate the quality process by asking would-be consultants two questions:

- What percentage of employees will be formally, structurally involved in the quality process? If the answer is anything less than 100 percent, ask who—specifically—is considered too stupid or recalcitrant to understand and support the proposal.
- How long will it take to launch a process that has a tangible impact on the bottom line? If your company has 3,000 or fewer employees, a quality process that includes both 100 percent involvement and process/value analysis (re-engineering without the attitude) can take as little as six to eight months. Consultants can help compress this time frame, as long as senior executives retain control and demonstrate leadership.

"What's next after quality?" moan the disillusioned. More quality—only done right this time. Finding the parts and pieces isn't hard; the resources listed are a good start. The trick is to get leadership, participation and measurement up and running at the same time. No consultant can do it for you.

On Achieving Excellence, *Oct. 1995.*

WHAT'S IN IT FOR ME?

"What's in it for me?" Far from being an idle query, this inquiry goes to the heart of what motivates a person to take part in a quality process. Discussing it as egoism or being downright selfish fails to recognize if for what it is: endearingly human.

No one, but no one, acts contrary to his or her own best interests, and the ability of the organization to answer the WIIFM (what's in it for me?) question is conceivably the best and most accurate predictor of how successful a quality process is going to be. Any top management team that wants to embark on a quality journey had better provide compelling answers, if it wants the crew to get on board for the voyage.

There's not one good answer, there's many... Fortunately, good answers are not hard to come by. If corporate executives think back on their own conversion to the TQM philosophy and discuss their experiences with management in their own and other companies, several things are likely to become evident:

1. Foremost among these is that everyone has a slightly different story to tell.
2. Not too surprisingly, different people are motivated by different ideas and by different goals. What gets one person excited will bore the second and barely amuse the third.
3. And that's okay. If the organization sticks with hiring humans, that kind of variability is part of the package.

The challenge: provide each with a door to quality... The challenge then is to present a broad spectrum of motivations and create an atmosphere in which it is possible for each individual to embrace the particular motivation, or combination of motivations, that he or she finds persuasive.

Which particular idea sparks which person is of little or no interest to the organization, so long as every employee is inspired to become an enthusiastic partner in the effort to continuously improve.

Consider the following: if the star of the local basketball team steps to the foul line with no time remaining on the clock, two free throws to shoot, and the team down by one point, nobody particularly cares why the player puts up the winning points. It could be because Mom is in the stands, because the championship depends on the results, or because he or she has an ego as big as a mid-western state. Everyone—fans, owners, team-mates—will be satisfied as long as the ball goes through the hoop twice in succession.

Country, company and me One reasonable clear, concise, and useful framework for answering what's in it for me breaks the benefits of quality into three categories: national, corporate, and personal. This allows the discussion to range over a wide variety of issues: macro versus micro, communal versus personal, tangible versus intangible, and rational versus emotional. How exactly does this work?

WHAT'S IN IT FOR US, NATIONALLY?

At the global level, as the nations of the world struggle to shake off worldwide recession, international competition is intensifying. There is growing recognition that if America is to control its national destiny, it must first control its economic destiny, and companies offering quality services and products help make that possible.

In 1988, the first Malcolm Baldrige National Quality Award was given to encourage companies with world class quality to share their experiences with corporate America. In that first year, only 12,000 applications were distributed; in 1992, over 200,000 applications were sent out. Clearly, more and more companies aspire to the role of quality leader.

Does this national motivation translate into individual motivation? It might and it might not. The clearer the connection between the actions of the individual and the well-being of the nation, the better the odds.

Globe Metallurgical's experience... The first Baldrige winner in the small business category (under 500 employees) was Globe Metallurgical. In the spring of 1987, it is highly improbable that many of the employees of Globe thought of themselves as national models or as frontline troops in the quality revolution. Since winning, employees have seen thousands of people come through their plant and have talked to representatives of hundreds of American firms of all sorts and sizes. As individuals, they know that their actions have a national impact—no doubt a powerful personal motivator for Globe employees.

The prospect of being a national player no doubt appeals to some. It can give an emotional dimension to work, transforming a *rationally not very exciting job* into something much more. On the other hand, for individuals whose power to make decisions is limited, the idea of being a national player may be too vague and/or too abstract to be of much interest. Perhaps, for them, knowledge of what the quality efforts does for an organization will be easier to digest—and more convincing.

WHAT'S IN IT FOR US, AS A COMPANY?

Quality makes money! Quality makes money!! Quality makes money!!! Companies that earn and maintain a reputation for being the quality alternative in their particular field are consistently profitable. Dozens of studies over the last decade provide data to support these assertions. The General Accounting Office management practices report *US Companies Improve Performance Through Quality Efforts* published in 1991 is particularly useful in this regard.

Rationally, that is a good news for employees... A secure company means secure jobs. A successful company means there is a bigger pie to slice when it comes time to determine wages, salaries and benefits. This monetary link may be the answer to WIIFM at its most basic level—although that is not the entire picture.

Emotionally, being part of a successful organization is a source of pride to a great many individuals. Those who look to community as part of the way they define themselves are going to find satisfaction in contributing to the success of an effort that adds to the strength of the organization.

Yeah, but I have to do more things, and differently... Yet for some employees, neither the potential impact on the future of the nation nor the company for which they work will induce them to change their daily behavior. The company is, after all, attempting to change the tacit terms of employment.

If I am currently producing at about 60 percent of my capabilities and the company has concurred up to that point that it expects no more, what's my incentive to change? I've been paid for that 60 percent; the company sends me a pay check at regular intervals and I accept it.

Maybe I don't enjoy the job much ... and maybe I don't even do a particularly good job, but both the company and I know where we stand. Now, with no talk of a pay boost, *they want something else*?

So, where's the clincher? Here's where motivation becomes strictly personal. Perhaps the biggest benefit the quality movement offers the individual is something that Dr. W. Edwards Deming stresses repeatedly:

The joy and dignity of honest work.

Deming and joy... A company that fully embraces Deming's quality concepts presents an emotionally rewarding work environment. In such an environment, workers are treated with respect, encouraged to participate in decision making, and have access to the information needed to do their work and make the enterprise successful. But beware of would-be Deming disciples who ignore the philosophy of joy and pride behind the numbers.

Control over your worklife... If a person only spends 40 hours a week at work and he or she sleeps seven hours a night, those hours at work represent almost a third of that person's conscious time. In return for learning to look at work in a new way, a quality process, well-implemented, offers each individual a greater degree of control over that rather substantial chunk of their life. By extending authority equal to responsibility throughout an organization, a quality process makes it possible for each person to be more of a player in the decisions that drive his or her day-to-day,hour-by-hour activities.

You can be you, not an it at work... Best of all, being involved in a quality process allows each individual to be the same person at work and at home. Too many American workers have been told, "Just work, don't think!" developing split personalities as a result. There is the competent, interesting individual known by friends and family, and there is the dullard whose brains have been left—as instructed—in the glove compartment of the car while he or she is at work.

A quality process asks, virtually begs, people to think and to be creative during working hours. For some people, it will take a little getting used to, but once hooked, they will never want to go back.

But I may not be here that long... Maybe an individual doesn't believe that his or her organization can live up to creating the environment described above. If someone is feeling pretty sure that he or she isn't going to stay at the current job much longer, is there still a good answer to WIIFM?

Yes, yes, yes. Quality is THE transferable job skill of the '90s. Keep in mind that one of the accepted truisms these days is that most Americans will be switching not only companies but whole careers with increasing frequency in the years to come. If a job-change is in your future, then participation in a quality process should be in your present.

They'r '90s kind of skills... Here's the hot skills shopping list for the 1990s:

- The habit of being open to continuous change and improvement
- The ability to talk with a customer and discover what his or her expectations are
- The willingness to take part in participative meetings and problem-solving sessions
- The technical knowledge of the vocabulary and tools of a quality effort

MASLOW AND QUALITY

This national–corporate–individual framework is fairly informal, but it parallels Maslow's hierarchy of needs, an accepted format for assessing and explaining human motivation. Found in many psychology textbooks, Maslow's theory states that all humans progress through the same sequence of needs, albeit at wildly varying speeds. Although hierarchical, the sequence is fairly untidy, depending on when each individual perceives his or her needs to have been met at one of five levels:

1. Physiological (survival)
2. Safety and security
3. Belongingness and love
4. Autonomy and self-esteem
5. Self-actualization

More like piano keys than stairsteps... It bears repeating that a person does not normally concentrate all energies on one need and then, when those needs are fully satisfied, move on to the next. At any one moment, a person may be concentrated mostly on one level (e.g., belongingness and love) but be simultaneously concerned to a lesser degree with the levels on either side of their primary target.

National government and basic survival structures

People traditionally look to a national structure to provide answers to physiological needs. Government itself may or may not be responsible for basic housing, clothing, and food—in short, survival—but it is responsible for providing the stability and infrastructure that enables these needs to be met. In Somalia, for instance, the local government's complete inability to cope has led to international intervention.

As with so many things in this area, success is relative. The steady increase in the income cut-off for poverty in America indicates that what Americans now define as *basic survival* is quite different from what was defined in this country in the past or in other countries now. Our government, at the behest of those we have elected over the past 50 years, has decided to step in with food stamps, ADC, Medicare and such when basic physiological needs are not met.

So, when individual employees see their contribution to a quality effort in national terms, it is (in part) a desire to insure that the nation can continue to be the guarantor of the perceived physiological needs of its citizens.

The company, safety, security, belongingness and love Bearing in mind that, overlapping with the national well-being, corporate benefits are most closely aligned with the second and third levels of Maslow's hierarchy.

Being an employee of a thriving organization goes a long way toward insuring the safety and security that comes from having a steady source of income. And being a member of a team can be a major component in satisfying an individual's need to belong and to be loved, whether the team is a 10-person work unit and/or an entire *corporate team of associates.*

The well run company, autonomy and self-esteem... Autonomy and self-esteem, the fourth level, have long been the backbone of a well-established quality process. What has been missing, perhaps, in so many failed or faltering efforts has been the realization that virtually no one can concentrate efforts in this area unless more basic needs are already being met.

That explains one of the paradoxes of the quality movement: the ideal time to begin a quality process is when the company seems not to need it. The chances of individuals within the organization reaching level four of Maslow's hierarchy are greatly improved when the company as a whole is healthy financially and emotionally. In those stellar cases where authority is pushed down the corporate ladder to where it is equal to responsibility, as in self-managed workteams, an increasing number of employees can achieve this level.

WIIFM and self-actualization Maslow's fifth level, self-actualization, is thought to be achieved by only a small percentage of individuals even under the best of conditions. Unless the surrounding environment insures that the individual is thought as a thinking, contributing adult and not just as another interchangeable part of the organization, reaching this level is virtually impossible.

That brings another paradox. The most accurate answer to the WIIFM question is this: When individuals take an active role in insuring the success of the quality effort, the company, now quality focused, will be able to create the environment that makes it possible for individuals to work their way through to a level of self-actualization at work. And along the way, the nation will benefit and the company will make a lot of money, but the gain to each person is the most dramatic and most precious.

The paradox... You can't get there unless you participate.

So perhaps the best answer to WIIFM is this:

WHAT'S IN IT FOR ME... IS A BETTER ME.

Journal for Quality and Participation, *March 1993*

CLOSING NOTES

The overlapping articles in this book, each written to highlight a particular aspect of quality, illustrate the point these authors made at the beginning: Quality is not complex. A quality process is made up of a few constant, relatively straightforward principles that anyone can grasp. Then, however, the real work of doing quality correctly begins. Designing a day-to-day process founded on sound principles requires careful thought; sustaining it after the design phase requires eternal vigilance and attention to detail.

Quality can be achieved by virtually any organization of any size and achieving it is provably worth the effort. Not only does quality bring benefits to the bottom line, it also improves everyone's quality of life. Quality enables everyone to make a contribution they can be proud of; it provides goods and services that consumers can rely on. And, in a world in which organizations are created and disappear, grow and shrink, with a frequency never seen before, quality is the transferable job skill of the 21st century. Equipped with a knowledge of quality theory and techniques, with the ability to work on teams, with the desire to continually improve, an individual becomes a more valuable employee no matter where he or she is employed.

Quality is not just the province of the CEOs or the gurus. Quality is everybody's business.

CLOSING NOTES

SOURCES

A list of all publications and articles:

American Press Review, Proceedings of Sixth Annual J. Montgomery Curtis Memorial Seminar; Quality Is Every body's Business

Best's Review; June 1987; The Policy Is Quality

June 1988; The Policy Is Still Quality.

Boardroom Reports, December 1, 1992, The Secrets of Continous Quality Improvement.

Distinguished Papers, St. John's University;

December 1989; The Service Revolution.

Executive Excellence;

February 1989; What Military Can Teach Business About Leadership

May 1989; Take It Personally

December 1994; Leadership at Every Level

July 1995; The Revolution Continues

The Journal for Quality and Participation

March 1988; A Quality Beginning

December 1988; Beginning "Quality Without Limits"

March 1989; Total Service Quality

June 1989; Try Continuous Involvement Improvement

January/February 1990; Breaking New Ground

March 1990; It Is Time To Get On With It!

September 1990; Quality–Down to the Roots

January/February 1991; Will Continuous Improvement Work Here?

March 1991; Warning: This Good Idea May Become a Fad

September 1991; Creating More Creativity

December 1991; An Excellent Enterprise If You Can Keep It

March 1993; What's In It For Me?

December 1993; The 100% Solution or Greater Nonconformity

March 1994; Beyond Charging the Bill and Demanding Excellence.

March 1995; A Revolutionary Example for Quality

December 1995; Work and Enjoyment

July/August 1996; The Importance of the Baldrige to US Economy

March 1997; Making Change Possible

Leader to Leader

Spring '97; The Three Priorities of Leadership

Managing Service Quality

Vol 5, #2; Quality Involves Everyone: How Paul Revere Discovered "Quality has Value"

Marine Corps Gazette

March 1993; Total Quality Leadership or Partial Quality Management?

Mass High Tech

Oct 10–23 1988; A New Model for Quality

On Achieving Excellence

August 1991; Pat Townsend on Choosing a Baldrige Quality-Assessment Consultant

December 1992; Are Your Practicing Total Quality? Take the Test

October 1995; What's Next After Quality?

Quality Data Processing

January 1989; The Right Question

Quality Digest (Internet)

November, 1997; What Went Wrong With Quality?

December 1997; Followership: An Essential Element of Leadership

February 1998; What Happened to Quality?

March 1998; Four Phases of a Quality Process

May 1998; Remaking a Quality Management System, Part One

June 1998; Remaking a Quality Management System, Part Two

July 1998; Top Management Commitment—What's That?

Quality Progress

June 1985; Insurance Firm Shows That Quality Has Value

July 1994; Sharing the Wealth in Quality Partnerships

January 1997; Qualicrats and Hypocrites: A Troubling Status Report from the Font

The Quality Observer

July 1992; Leadership: An Ancient Source for a Modern World

September 1992; Participation: Starting with the Right Question

October 1992; Measurement: Neither a Religion nor a Weapon.